T0207433

Dimensionality Reduction in Data Science

Max Garzon •
Ching-Chi Yang • Deepak Venugopal •
Nirman Kumar • Kalidas Jana • Lih-Yuan Deng
Editors

Dimensionality Reduction in Data Science

Springer

Editors
Max Garzon (iD)
375 Dunn Hall
The University of Memphis
Memphis, TN, USA

Deepak Venugopal
375 Dunn Hall
The University of Memphis
Memphis, TN, USA

Kalidas Jana
Memphis, TN, USA

Ching-Chi Yang (iD)
373 Dunn Hall
The University of Memphis
Memphis, TN, USA

Nirman Kumar (iD)
375 Dunn Hall
The University of Memphis
Memphis, TN, USA

Lih-Yuan Deng (iD)
373 Dunn Hall
The University of Memphis
Memphis, TN, USA

ISBN 978-3-031-05373-3   ISBN 978-3-031-05371-9   (eBook)
https://doi.org/10.1007/978-3-031-05371-9

This Springer imprint is published by the registered company Springer Nature Switzerland AG
The registered company address is: Gewerbestrasse 11, 6330 Cham, Switzerland

Preface

Data science is about solving problems based on observations and data collected in the real world. Problems may range from the mundane to difficult scientific questions, for example, rating movies for recommendation systems, understanding the earning power of American taxpayers, increasing revenue for a business, spam, controlling the spread of misinformation through the internet, global warming, or the expansion of our universe. Our ability to generate, gather, and store volumes of data in the order of tera- and exo-bytes daily has far outpaced our ability to derive useful information from it in many fields, with available computational resources. The overarching goal of this book is to provide a practical and fairly complete, but not encyclopedic, review of Data Science (DS) through the lens of Dimensionality Reduction (DR).

The intended audience consists of professionals and/or students in any domain science who need to solve problems to answer questions about their domain based on data. Domain science is a fairly vague term that refers to a specialized area of human knowledge characterized by specific questions about a certain aspect of reality (like what is motion in *physics*, what are physical objects made of in *chemistry*, what is life in *biology*, and so forth.) In addition to the well-established sciences, they include just about any area where data can be recorded and analyzed to answer questions concerning the population of individuals or objects the data is about.

Data science presents a singular approach to problem solving when compared to more established sciences. Traditional sciences are motivated by understanding our world in order to survive and thrive. That requires a degree of analysis and theorizing to understand the specific phenomena involved and to enable predictive power. By contrast, with the advent of computer science and its abstractions into the information age (as embodied by the internet and web for example), tools have been created that can be used regardless of the specific domain. Once this threshold is crossed, then it is a natural next step from mathematics and statistics to synergistically combine them with the powerful computational tools developed by computer science to create a new science that is more than the sum of the parts, hence data science.

We have strived to leave our niche hats at the door and present an intuitive, integrative and synergistic approach that captures the best of the three worlds. That is the pervasive thread that readers will discover through examples and methods throughout the book. Sections begin with intuitive examples of a problem to be solved by the (perhaps new) concepts and results being described in the section. A professional with an undergraduate degree in any science, particularly quantitative, should be able to easily follow this part. These motivating examples are then followed by precise definitions of the technical concepts and presentation of the results in general situations. These require a degree of abstraction that can be followed by re-interpreting concepts like in the original example(s). Finally, each section closes with solutions to the original problem(s) afforded by these techniques, perhaps in various ways to compare and contrast dis/advantages of the various DR techniques based on quantitative and qualitative assessments back in the real world.

We are grateful to acknowledge support for this project from various sources. First, support from the University of Memphis CoRS (Communities of Research Scholars) program (through Deborah Hernandez) to start up research projects that facilitated initial interactions that eventually led to the interdisciplinary collaboration that produced this book, among other works, as well as access to the High Performance Cluster (HPC) used in development and testing of most results herein. Second, Lih-Yuan acknowledges support from the National Science Foundation under an award for The Learner Data Institute (NSF-1934745: The opinions, findings, and results are solely the authors' and do not reflect those of the funding agency.). Third, Max acknowledges faculty (Professor Luis Fernando Nino) and student support (especially Alfredo Bayuelo for TA support and help with cartoon design and rendering) at the National University of Colombia for their active feedback and project outcomes at the XI International Cathedra on *Data Science for Bioinformatics* in 2017, where several of the approaches, views, and some results included in this book were conceived.

Bon voyage!

Memphis, TN, USA   Max Garzon
November 2021   Ching-Chi Yang
                                                                           Deepak Venugopal
                                                                            Nirman Kumar
                                                                             Kalidas Jana
                                                                            Lih-Yuan Deng

Contents

Contents

Acronyms

General

DS, DR, ML	Data Science, Dimensionality Reduction, Machine Learning
P	Probability (measure)
p, n	Dimensionality p, size n of a dataset

Data Science

[IrisC]	Iris Flower Classification problem
[MalC]	Malware Classification problem
[BioTC]	Biotaxonomy Classification problem
[SynC]	Noisy Synoptic Data Classification problem
[AstroP]	Planetary Body Location Prediction problem
[PhenP]	Phenotypic Prediction problem
[RossP]	Rosette Area Prediction problem (of a plant)
[LocP]	Provenance of a Plant Prediction problem
MDS	Multi-Dimensional Scaling (classical or metric)
ISOMAP	Isometric Mapping
t-SNE	t-Stochastic neighbor embedding
RND	Random Scaling
MNIST	MNIST dataset

Statistical

Ω	A population/sample space
D, X, Y	A sample or dataset from Ω
P	Probability (measure)
$P(B \mid A)$	Conditional Probability of event B given event A
$X, Y, \cdots$	Random variables (RVs, possibly with subindices)
$E(X)$	Expected value (mean) of a RV X
$Var(X)$, $\sigma(X)$	Variance, standard deviation of a RV X
H	Shannon Entropy

N(0,1)	Standard Normal Distribution
N(μ, σ)	Normal Distribution with mean μ and standard deviation σ
EDA	Exploratory Data Analysis
PC, IC	Principal Component, Independent Component
PCA, ICA	Principal, Independent Component Analysis
KPCA	Kernel Principal Component Analysis
LDA,QDA	Linear, Quadratic Discriminant Analysis
GLM	Generalized Linear Model
MLR	MultiLinear Regression
BL	Bayesian Learning
SVD	Singular Value Decomposition
SVM	Support Vector Machine

Computational/Machine Learning

AP	Computational/Algorithmic problem
[**CharRC**]	Handwritten Character Recognition problem
[**TSP**]	Traveling Salesperson problem
$\mathbf{D}_m$	The DNA space of Watson-Crick paired oligomers (pmers) of length m
$h, h(x, y), \lvert *, *\rvert$	Hybridization h-distance in $\mathbf{D}_m$
E, P^i	Equator, i^{th} Parallel in DNA space $\mathbf{D}_m$ (possibly with subindices m)
[**CWD**]	Noncrosshybridizing (nxh) basis/microarray Codeword Design problem
GNB	Gaussian Naive Bayes
MLR	MultiLinear Regression
MLP	MultiLayer Perceptron
FFN, NN, CNN	Feed-Foward, Neural, Convolutational Neural Network
SVM	Support Vector Machine
RF	Random Forest
kNN	k-Nearest Neighbors
NFL	No Free Lunch Theorem
R^2, AICS, BIC	Assessment metrics of statistical solutions: R-squared, Akaike's Information Criterion, Bayesian Information Criterion
$a, p, r, F1, RMSE, RE, s$	Assessment metrics of ML solutions: accuracy, precision, recall, F1-score, Root Mean Square Error, Relative Error, silhouette

Mathematical

$\approx, \propto, \ll, \gg$	Approximately equal to, proportional to, much smaller/greater than
$\alpha, \beta, \lambda, \nu, \sigma, x, y, \cdots$	Scalars (possibly with subindices or superindices)
$\mathbf{x}, \mathbf{y}, \mathbf{u}, \mathbf{v}, X, Y, \ldots$	Vectors (possibly with subindices or superindices)
$\mathbf{C}, \mathbf{X}, \mathbf{Y}, \mathbf{W}, \cdots$	Matrices (possibly with subindices)
$\mathbf{N}, \mathbf{Z}$	The set of natural numbers $\{0, 1, 2, \cdots\}$, integer $\{0, \pm 1, \pm 2, \cdots\}$ numbers
$\mathbf{Q}$	The set of rational numbers $\{p/q : p, q \in \mathbf{Z}, q \neq 0\}$
$\mathbf{R}$	The set of real numbers $(-\infty, +\infty)$
$\mathbf{D}_m$	The DNA space of Watson-Crick paired oligomers (pmers) of length m
$|*, *|; h, h(x, y), |xy|$	Distance; Hybridization h-distance in $\mathbf{D}_m$
NMF	Nonnegative Matrix Factorization

Chapter 1
What Is Data Science (DS)?

Max Garzon ⓘ, Ching-Chi Yang ⓘ, and Lih-Yuan Deng ⓘ

Abstract Our ability to generate, gather, and store volumes of data (order of tera- and exo-bytes (10^{12}–10^{18} bytes) daily) has far outpaced our ability to derive useful information from it in many fields, with available computational resources. The theme of this book is a review of Data Science (DS) through the lens of Dimensionality reduction (DR). Data science is about solving problems based on observations of factors (referred to as *co-variates*, *predictors*, or just *features*) that may determine a solution. Typical kinds of problems are described, including classification, prediction, and clustering problems, as well as data collection methods.

1.1 Major Families of Data Science Problems

In this section, problems typically dealt with in data science are described and grouped into three major groups: **classification, prediction,** and **clustering**.

A *population* Ω is the entire group of individuals/objects of interest for a given problem, while a *sample* refers to a (relatively very small) subset $\mathbf{D} \subseteq \Omega$ of the population.

Data science problems can be regarded as computational problems in the sense that they call for certain types of inputs as data to a problem (as described in Sect. 1.3 below) and spell out an expectation in terms of the kind of result to be produced as a solution. The solution is usually some sort of model or program/code, ultimately to be translated into some sort of physical device to be run, typically a conventional computer. In computer science, these problems are referred to as *algorithmic problems* and are typically expressed in the forms of questions of a

M. Garzon (✉)
Computer Science, The University of Memphis, Memphis, TN, USA
e-mail: mgarzon@memphis.edu

C.-C. Yang · L.-Y. Deng
Mathematical Sciences, The University of Memphis, Memphis, TN, USA
e-mail: cyang3@memphis.edu; lihdeng@memphis.edu

similar kind being provided as inputs from a population of objects, as illustrated with the three major families described in the following subsections.

1.1.1 Classification Problems

One of the most common kind of problems is *classification*.

Example 1.1 Iris flower classification is a relatively simple example. The population of objects under consideration for inputs is a feature vector x giving the length and width of sepals and petals of an iris flower. The flower is to be placed in one of the three categories. These categories must be specified in advance for the problem and determine the type of classification being made into disjoint and exhaustive classes, i.e., they constitute a *partition* Π of the population. If the categories change, then we are dealing with a different problem since the answers will have to change across the inputs. The problem can be stated precisely as follows, with $\Pi_1 = \{Setosa, Versicolor, Virginica\}$. □

[IrisC]   IRIS FLOWER CLASSIFICATION (Π_1)

INSTANCE: A feature vector $x =$ (sepal length, sepal width, petal length, petal width) (in cms) describing an iris flower, e.g., $(4.6, 3.1, 1.5, 0.2)$
QUESTION: Which kind of flower in Π_1 is x?

The brackets contain a mnemonic name **[IrisC]** used to refer to the problem in the sequel. A dataset for this problem is described in more detail in Sect. 11.5.

Example 1.2 Malware classification is a more complex example. The population of objects under consideration for inputs is a computer program/code that causes harm when run on a computer, usually referred to as *malware*. A sample from this population is discussed above in Sect. 1.1, where a given piece of malware is expected to be categorized into nine (9) classes or categories, as shown in Table 1.1. □

The problem can then be stated precisely as follows.

[MalC]   MALWARE CLASSIFICATION (Π_2)

INSTANCE: A piece of malware x
QUESTION: Which category c in Π_2 does x belong in?

The Malware Classification Challenge (*arxiv.org/abs/1802.10135*, accessed October 2021) dataset was released in 2015 and is publicly available through *kaggle.com*. An available dataset consists of a set of 10,868 known malware files representing a mix of 9 different malware families, as summarized in Table 1.1. Each datapoint contains both the raw binary content of the malware file as well as metadata information extracted using the IDA disassembler tool, for a total of

Table 1.1 Microsoft's malware classification problem. The labels for the classes in the partition Π are in the first column

Labels	Class name	Type of malware	Count
1	Ramnit	Worm	1541
2	Lollipop	Adware	2478
3	Kelihos_ver3	Backdoor	2942
4	Vundo	Trojan	475
5	Simda	Backdoor	42
6	Tracur	TrojanDownloader	751
7	Kelihos_ver1	Backdoor	398
8	Obfuscator.ACY	Any kind of obfuscated malware	1228
9	Gatak	Backdoor	1013
			10,868

1804 raw features. The classification challenge is to correctly assign every piece of malware in the population to one of the nine (9) possible malware classes defined with the challenge. A dataset for **[MalC]** is described in more detail in Sect. 11.5.

Example 1.3 A third and more difficult example of a classification problem is species identification in biology for a given group of species (a taxon) T. The population of objects under consideration consists of all living organisms belonging to a species in T. The taxon T is the union of all specimens/individuals in the species in the second column of Table 1.2, according to a given biological taxonomy, while the partition labels Π are shown in the first column.   □

For a data sample, representative genes were obtained from mitochondrial DNA (specifically, cytochrome Oxidase genes COI, COII, and COIII, CytB) of 249 organisms (as shown in the fourth column of Table 1.2) collected from biological genetic repositories (e.g., GenBank at *www.ncbi.nlm.nih.gov/genbank/*). A dataset for this problem is described in more detail in Sect. 11.5. The problem can then be stated precisely as follows for an arbitrary taxon T.

[BioTC] **BIOTAXONOMIC CLASSIFICATION**

INSTANCE: A DNA sequence representing a living organism x
QUESTION: What species in T does x belong in?

In summary, a *classification problem* is defined by a partition Π of a population Ω (into a number of *mutually disjoint classes exhaustive of* Ω) and the problem is to place an arbitrary element of the population into the right class. A solution to the problem is a model that will make a (hopefully correct) decision for every element from the population (*not just the sample*) as to which class it belongs to.

Table 1.2 Organisms in the sample data for **[BioTC]** of a biological taxon. The 21 classes/labels in the partition Π are in the first column

Label	T: genus species	Common name	Count
1	Apis mellifera	Western honey bee	4
2	Arabidopsis thaliana	Cress	5
3	Bacillus subtilis	Hay/grass bacillus	18
4	Branchiostoma floridae	Florida lancelet	18
5	Caenorhabditis elegans	Round worm	6
6	Canis lupus	Wolf	18
7	Cavia porcellus	Pork	4
8	Danio rerio	Zebra fish	9
9	Drosophila melanogaster	Fruit fly	18
10	Gallus gallus	Red junglefowl	18
11	Heterocephalus glaber	Naked mole rat	3
12	Homo sapiens	Human	18
13	Macaca mulatta	Rhesus macaque	8
14	Mus musculus	House mouse	18
15	Neurospora crassa	Red bread mold	5
16	Oryza sativa	Asian rice	12
17	Pseudomonas fluorescens	Infectious bacterium	6
18	Rattus norvegicus	Brown rat	18
19	Rickettsia rickettsii	Tick-born bacterium	18
20	Saccharomyces cerevisiae	Yeast	18
21	Zea mays	Corn/maize	7
			249

1.1.2 Prediction Problems

A second kind of problems is *prediction*.

Example 1.4 The population could be plants x (e.g., a specimen of *A. thaliana*) and the question of interest is to know the area of the rosette of the plant covered by the spread of its leaves at maturity. The problem can then be precisely defined as follows for a taxon of plants T.

[RossP] **ROSETTE AREA OF A PLANT** (T)

INSTANCE: a (long) DNA sequence x (over the alphabet $\{a, c, g, t\}$) representative of a plant in T?

QUESTION: What is the area of x's rosette when fully grown?

Note that there is a dependency (not necessarily functional since environmental factors appear to be involved, e.g., how well the plant was fed) that roughly determines the rosette area given the genome of the plant. A similar problem can be posed for the weight of a person as a function of their height. □

Example 1.5 Another example is the problem [**LocP**] of determining the origin of the plant for a taxon of plants T based on DNA sequence alone, as stated below. □

[**LocP**] PROVENANCE OF A PLANT (T)

INSTANCE: a (long) DNA sequence x (over the alphabet $\{a, c, g, t\}$) describing a plant in T

QUESTION: Where on Earth (latitude, longitude) was x grown?

In summary, a *prediction* problem is defined by a function f on a population that associates numerical values to every element in the population (usually hard to measure directly) and the problem is to find the value $f(x)$ of the function f for an arbitrary element x in the population Ω. A solution is a model to determine that value based on other features identifying the input elements x from Ω. Since this may be difficult, one may need to settle for just approximate values.

1.1.3 Clustering Problems

A third kind of common problem is *clustering*.

Example 1.6 When a typical US town grows to a certain size, it is desirable to set up some fire stations just in case a house or building catches fire. Naturally, when your house is on fire, you would like the station to be as close as possible. To conceptualize the problem, we can specify a number n of points c_i $(i = 1, \cdots, m)$ in a 2D Cartesian plane to designate the positions of the fire stations. Thus, any fire emergency at a house located at any given position x will be tended to by fire trucks at station c_{i_0} closest (at minimum distance) to it in distance, i.e., so that

$$i_0 = arg \min_i \|c_i - x\|.$$

So, if there are two stations c_1, c_2, locations get clustered in two clusters, defined by half-planes separated by the perpendicular bisector of the segment joining them. If there are three stations, the regions are like cake slices, as shown in Fig. 1.1a, with boundaries being the relevant fragments of the perpendicular bisectors of all three pairs of stations. More stations will complicate the partition of the 2D plane, as illustrated in Fig. 1.1b, but along the same ideas. This kind of diagram is called a *Voronoi diagram* and is defined by the fire station locations $c_1, \cdots, c_m$, which are generically referred to as *centroids*. This diagram is a solution to the problem of clustering the population into groups that are each served by the same fire station according to the ordinary Euclidean (ℓ_2) distance. □

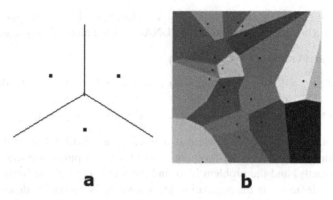

Fig. 1.1 Voronoi diagrams as solutions to the clustering problem defined by the Euclidean distance in the population of the 2D plane for (**a**) three points (e.g., fire stations) and (**b**) 20 points

In summary, a *clustering problem* is defined by a metric (usually a distance function, defined precisely below) function between elements in a population to capture degrees of (dis)similarity, and the problem is to produce a partition (*unlike a classification problem, where the partition is given*). A solution is to find a model that will produce a partition such that elements in any one cluster are more similar among each other than to elements in the other clusters.

It may be difficult to make a crisp distinction between these kinds of problems. For example, a partition Π in a given classification problem can be regarded as a function f on the same population assigning a numerical label in an enumeration of the classes in some order, so it would appear that every classification problem is a prediction problem (especially if a solution to the problem is used to label unknown inputs). Conversely, if the range of f in a prediction problem is small, one can regard it as a classification problem as well. Thus, we will speak of prediction problems only when the range of f is a fairly large set of (perhaps infinitely many) analog values. Therefore, typical classification problems usually involve a few categories in their partitions (in the order of at most tens). Likewise, once a solution to a clustering problem is obtained, one can use the clusters as a partition in a classification problem to predict the clusters for unknown elements in the population.

Other kinds of problem besides these three categories can certainly arise in data science. An example involves dynamical sequences of actions as inputs (e.g., a video of a pedestrian crossing a street) and questions concerning the next expected action (e.g., will the pedestrian run across or turn back on an upcoming car). There is no telling what kind of problem could arise in any domain science. However, these three main categories can always be used, at least in the first approximation to a given problem. For example, the answers to the pedestrian problem can range from a simple classification (forward/stop/back) to a prediction of the next step, to prediction of complex sequence of moves. (What sequence of actions will the pedestrian exactly take in the next 5 seconds?)

1.2 Data, Big Data, and Pre-processing

In this section, a characterization of what is understood by data and major techniques to conceptualize, visualize, and understand data are reviewed. Understanding data, particularly large datasets and big data, can prove to be extremely difficult, particularly when faced with acting on decisions based on data analytics discussed in later chapters to sell a solution of a particular problem to management. Further, the results of data analysis may impact an individual's health, or even large communities when applied at large scales in the real world, so understanding the scope and rationale for the results of the analyses is critical.

1.2.1 What Is Data?

A basic question that any book on data science must address is, what exactly is *data*? Many answers have been given to this question since the dawn of the information age in the 1940s. For example, well-known computer scientists (such as Peter Naur of BNF: Backus-Naur Normal Form fame) wanted to redefine computer science as *data science* back in the mid-1950s. So have well know statisticians (such as C.F. Jeff Wu of EM optimization and '*Statistics = Data Science*' fame in the 1970s), implying that data is about observations in the form of numbers, that it can be used to make inferences about entire populations and that there is a computational component to it. Like an iron age man on attempting to answer the question *What is iron?*, we can tell data when we see it but are very hard pressed when asked to give a rigorous noncircular definition of data. Plain numbers in abstract or computer programs to crunch them cannot really be data as it is understood today. We will thus adopt the following characterization.

> *Data* is an objective *recording* of one or several event(s) in the real world that is accessible at later times for perusal and analysis by at least one person.

Example 1.7 Thus, the following do *not* qualify as data: the plain occurrence of an event; a memory in someone's head as a witness of an event (unless an objective recording is made of the witness describing or reliving the event -humans update their memories as they remember them); the contents of a computer's RAM memory (unless it is being recorded); and so forth. On the other hand, the following can be considered data: website contents containing inaccurate facts and spreading misinformation; inaccurate scientific observations (due to faulty equipment or even malicious intent); fossils of dead or extinct animals; videos of a person (even if no longer living); and statistics collected through observations or machines and stored on paper or machine. Thus, data requires the so-called four Ws: *What, Who from, Where, and When.*   □

It will be useful to make further distinctions with other related concepts, such as *information* and *knowledge* (*wisdom* is out of reach). In this book, data differs from information in that information requires, in addition, a purposeful observer, for example, trying to solve a (data science) problem, as discussed below in Sect. 1.3. Data becomes information if it can be used to extract a solution to the problem with some processing, as illustrated by many examples in the following chapters. (The more complex concept of knowledge will be re-visited in Sect. 10.2).

Data can assume a variety of formats, for example, numbers, text, DNA or protein sequences, voice recordings, images, x-rays, videos, browser histories, time stamps, and so forth. These recordings are considered to be *raw* or *primary* data. They usually require transformation and processing to enable the kind of analysis done in data science to solve a problem, as will be discussed in this section and later chapters.

The most common format for the simplest kind of data is a two-dimensional (henceforth, just 2D) *table* consisting of rows and columns, usually organized so that attributes of the data are placed in the columns and each specific observation (sometimes referred to as a *datum*) occupies an entry in the rows. Herein, they will be referred to as *features* and *records*, respectively. Therefore,

> a *record* is a p-dimensional (henceforth pD) vector of features values (datums, or its latin plural, *data*). A table can be regarded as an nD vector of pD vectors, the records. The *dimensionality* (*size*) of the data/sample is p (n, respectively.)

Mathematically, a table can be regarded as a sample and its columns can be regarded as random variables X, as described above in Sect. 1.1 (see Sect. 11.1 for statistical background concepts). Single tables are good enough for small datasets but a more complex dataset or problem may require many such tables to form a *data corpus* for the problem simply because there may be many very different kinds of observations and attributes about objects in the same population the corpus is all about. Data corpora raise the problem of properly organizing the data so that humans can comprehend it. A criterion or logic to *organize* a dataset or data corpus is usually called an *ontology* for the data.

Example 1.8 If the problem of interest is the earning power of American taxpayers, the underlying population is, of course, American citizens who must pay income taxes by law. An important table here would contain a taxPayerID, a fiscal year, an AGI (AdjustedGrossIncome), and possibly taxDue. Another table may include demographic information, including taxPayerID, taxPayerFullName, CurrentAddress, priorAddress, and bankAccountNr as features. A third table may contain information about itemizedDeductions for a given year, and so forth. Many other tables are probably necessary to keep a full history of a taxPayerID in the full data corpus. The ontology in this case is the grouping of the information by taxPayerID and into a number of categories (like tax data for the first table, demographic data for the second, itemized deductions for the third) for each fiscal year. Other ontologies are certainly possible (e.g., brute force, a single table with one record for each American tax payer, and hundreds or even thousands of attributes for each of them), although probably less desirable because it is less useful to work with and/or understand the data.   □

On closer examination, a feature and its values (datums) in a table can be classified in various ways. For example, they could be qualitative or quantitative variables. A *qualitative* variable only observes values in a limited and fixed set of possible outcomes. If the list of outcomes can be sorted naturally, they are referred to as *ordinal*. Otherwise, the feature is called *nominal*. On the other hand, a *quantitative* variable observes values with a very large number of possibilities, usually the result of measuring something. They can be further separated into *discrete* or *continuous* variables. A discrete variable takes a value from a set of finite or countable numbers (indexable by natural numbers), while a continuous variable takes values from a set of uncountable numbers (like real numbers).

Example 1.9 (Qualitative or Quantitative) In order to distinguish qualitative and quantitative variables, a useful criterion is, *Can the values of the variable be added?* If a variable represents what is in a shopping bag, the bag may contain *oranges* and *apples* as answers. Since oranges and apples cannot be added (orange-apple makes no sense as a fruit), the list of items in the bag is qualitative data. However, if one is counting items in the bag, the data will be quantitative, say 2+3=5 items. □

Example 1.10 (Ordinal or Nominal) It is hard to objectively decide on a natural order for apples, oranges, and bananas (though their *names* can easily be sorted alphabetically). The variable typeOfFruit is nominal. On the other hand, a Likert scale for answering whether one likes a particular fruit is ordinal (e.g., *Like, Neutral, Dislike*). □

Example 1.11 (Discrete or Continuous) The criterion for distinguishing discrete or continuous variables can be, *Could a decimal number be an answer?* For example, if one is counting how many people are in a theater, a number such as 3.14 is nonsense. The number of people in a theater is discrete. However, if one considers the average number of people in a theater for one show over a period of one year, the number 3.14 could be an answer; this variable is continuous. □

1.2.2 Big Data

A similar problem to that in Example 1.8 arises with financial credit card transactions for a given bank or credit card company and many other problems for government and private organizations in every country. The data corpora can become really huge and they are usually referred to as *big data* (in the order of terabytes and larger). This fact raises the important issue of *understanding* datasets and data corpora. A good ontology will enable a human to grasp the large picture of what the data is all about (the four Ws) and make it possible to develop effective strategies to solve problems about the given population.

Big data is usually characterized by the so-called five V's:

- *Volume* (raw number of records/amount of data),
- *Variety* (how diverse is the type, nature, and format of the data),

- *Velocity* (speed of data generation),
- *Veracity* (trustworthiness/quality of captured data), and
- *Value* (insight and impact afforded by the data about the overall population).

Example 1.12 (Mobile Data with High Volume and Velocity) According to *ericsson.com*, total global mobile data traffic reached 49 exo-bytes (EB) per month at the end of 2020. It is expected to grow with the launch of the 5G internet. By comparison, the Iris dataset only contains 2.6×10^3 bytes, whereas a month of total global mobile data contains 4.9×10^{18} bytes. The data is generated extremely fast *www.ericsson.com/en/mobility-report/dataforecasts/mobile-traffic-forecast*, so this dataset exhibits very high volume and velocity.   □

Example 1.13 (Data Variety) Long ago, data might have referred to a well-organized spreadsheet, e.g., .txt or .csv files. Variety refers to any data that could be mixedly generated by humans or by machines. For example, a Fitbit generates sensor data of a body by continuously monitoring the body movements. Browser history is generated by human activity. Twitch stream is a live video that records the show and chats between hosts and viewers. A data corpus containing these chats has high variety.   □

Example 1.14 (Anonymity on the Internet) For freedom of speech, websites may offer online anonymity. However, it comes with the problem of data veracity. It is easy to lie and the information therein cannot be trusted. Moreover, fake information usually spreads faster than facts.   □

Example 1.15 (Educational Data) Data value is a subtle concept. It refers to how useful the data is to answer research questions and turn it into business intelligence. In the case of educational data, students' progress/scores are stored from K-12th grades. If the data are just stored without further use or analysis, it offers little value. However, if one can utilize the data to further identify personalized key learning strategies for students, its value increases tremendously for the benefit of society.

□

Computational resources and techniques today are not able to process the extraordinary full volume of data being generated in so many fields. Because of the overwhelming growth of data collection, it can be challenging to extract useful information from big data with current computational resources. Choosing a good subset of the data as the training sample to build some suitable models is a critical issue. For proper big data modeling and analysis, it is common to use some statistical and machine learning methods for preliminary pre-processing in order to make big data amenable to analysis. It helps to reduce the computational time of the analysis from years to feasible.

1.2.3   Data Cleansing

Raw data, i.e., data that has not been processed and is not ready for analysis, requires *pre-processing* or *cleansing* (e.g., (re)formatting, organization, and selective extraction) to become data ready-to-use (sometimes referred to as *cooked data*). The process is also referred to as data *cleaning, scrubbing, wrangling*, and so forth. Cleansing can be manually done or automatically done by a machine for large datasets.

Example 1.16 (Handwritten Character Recognition) In handwritten character recognition, a trace of handwriting is processed (e.g., zip codes on ordinary mail envelopes in order to route the envelopes properly). Further details about this dataset can be found in Sect. 11.5.   □

Example 1.17 (Malware Detection) Malware Classification ([**MalC**]) requires professional/domain knowledge to extract the information from an executable (.exe file) into cleanly labeled data suitable for analysis, where the rows are the pieces of observed malware and columns are the features extracted from the executable file.
                                                                        □

Example 1.18 (Stock Market) In the stock market, a trade can be made and recorded nowadays in approximately $\frac{1}{64}$ millionth of a second, i.e., at the rate of 64M per second. It is difficult to analyze this raw data because of its volume. It has to be organized and transformed for further analysis/prediction, e.g., by trading volume or opening/closing price in a day, to be of value.   □

In Examples 1.16–1.18, raw data are not ready for answering research questions. Additional procedures are required to organize observations and their features.

Cleansing is the process of screening and/or transforming data to enable processing and analysis to derive value from it.

Needless to say, cleansing must be conducted prior to data analysis, although it is sometimes done later upon failure to get useful results from an analysis due to multiple reasons. The data may be *incomplete, noisy, inconsistent, duplicated, irrelevant*, or simply be *missing* values. The aim of cleansing and pre-processing is to remove noise, standardize the format, and retain useful information, so as to make sure the scripts for analyses that expect a common format will not crash due to formatting errors or missing values. This is a pre-condition to extracting value from the data analyses. There are several common ways to accomplish this goal.

1.2.3.1   Duplication

For example, if a minority opinion is surveyed repeatedly and a large number of duplicates are added to the dataset, the minority opinion might become the majority opinion. By removing duplication, one can avoid biased inferences and misleading conclusions.

Example 1.19 (Online Sweepstakes) Online sweepstakes offer great prizes. If one enters multiple times, the case is duplicated. To be fair, the website keeper should remove the duplication or prevent a person from entering the sweepstakes twice. □

1.2.3.2 Fixing/Removing Errors

It is always recommended to re-check the source to obtain the correct values for the missing values. However, this is not always possible. Certain values cannot be recovered in many situations. One way to handle it is to change the erroneous datum to a missing one and handle it accordingly.

1.2.3.3 Missing Data

Missing data/value refers to a datum being unavailable or corrupted. Data can be missing due to many reasons. For example, in a survey, an individual chose not to report an answer; the datum was not observable/available; or the datum went missing during data (pre-)processing. *Missing value imputation* is one common technique to obtain a missing value. Some common imputation methods are:

1. *Mean imputation*
   The missing value x_{ij} for the variable X_j can be filled in as the mean, i.e., $\bar{X}_j$. This method is generally used for a continuous variable since the mean of a discrete variable or a categorical variable might not make sense or be appropriate.
2. *Interpolation and extrapolation* (for numerical data)
   The missing value can be imputed by prediction. While interpolation implies the prediction falls within the range of data points, extrapolation implies the prediction falls outside the range of data points. Basically, by connecting two close data points without missing values with a straight line, the missing values can be imputed as the value on the fitted line.
3. *Regression imputation*
   The missing value can also be imputed by a model for the data obtained later by analysis. One type of model is the regression type, which will be discussed in detail in Sect. 2.1.

1.2.3.4 Outliers

An *outlier* is a data point containing a datum that is out of the range of the other points in the feature. While the multivariate situation is considered, a distance measurement, e.g., Mahalanobis distance, will be used to disrobe how far the observation is from the bulk of the data. The goal of a whole research area called *outlier detection* (also *anomaly detection*) is to estimate this distance and identify the outliers. It can be used for detecting unusual observations, such as bank fraud.

Example 1.20 (Bank Fraud Detection) It is essential to detect fraudulent transactions not only for a personal account but also for money laundering. If an account usually has transactions for grocery shopping, the occurrence of a large transaction appears as an outlier. A bank will issue a hold on this type of activity until confirmed. How large is considered "large" requires fine-tuning. (While keeping its customers' assets safe is important, the bank will not want to issue a hold every time and ruin most honest users' experience.)   □

1.2.3.5   Multicollinearity

Multicollinearity (also collinearity) refers to the case where one predictor variable in a multiple regression model (described in Sect. 2.1) can be linearly predicted from the others with high accuracy. This technique is similar to combined inter-extrapolation, but with more than one missing value. (Some consider dependencies of this type to be "dirty" and require cleansing.)

1.2.4   Data Visualization

One of the major problems with data, particularly big or complex data, is to make it *understandable* to humans. This is particularly important for humans to verify, interpret, and/or rationalize the results of any analytics to act on a solution to a problem as being consistent with the data that led to it. In Human–Computer Interaction (HCI), the problem is usually solved by reaching into the human's head and finding a metaphor to transform and present the data to enable his/her brainpower with an appropriate transformation into a more familiar situation.

Example 1.21 A typical example is the problem of navigation on the metro system in a big city. A computer system can only store a list of metro lines and the locations (latitude or longitude) of their stop stations. Presenting a time table of these stations to a human will not enable easy navigation to instruct her to move from point A to point B. The visualization in the form of a graph in various colors (the lines) roughly following the layout of the city and their points of intersections and stops along the lines is a familiar and effective solution because humans easily understand space, distances, and orientation to navigate the city.   □

In statistics, data visualization refers to exploratory data analysis (EDA) as an approach to analyzing datasets to summarize their main characteristics, often using graphics about statistics of the data features. Based on the variable's type (qualitative/quantitative, discrete/continuous), different EDA methods can be used. Table 1.3 and Figs. 1.2, 1.3, 1.4, 1.5, and 1.6 illustrate various techniques with the Iris dataset.

Table 1.3 A stem-and-leaf plot of the sepal length for the Iris dataset

Key: 4l3 = 4.3 cm
4 l 3444
4 l 566667788888999999
5 l 0000000000011111111122223444444
5 l 555555566666677777777788888888999
6 l 0000001111112222333333333344444444
6 l 55555566777777778889999
7 l 0122234
7 l 677779

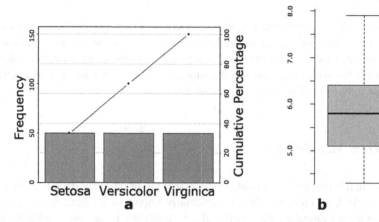

Fig. 1.2 (a) A Pareto chart with the count for the species and (b) a boxplot for the sepal length of the species in the species in the Iris dataset

Example 1.22 (Boxplot of a Continuous Variable) Boxplots are useful to represent the five number summary of a feature: the *minimum*, the *first quartile*, the *second quartile*, the *third quartile*, and the *maximal* of a variable. Figure 1.2b shows a boxplot of the sepal length for the Iris dataset.

Note that *quantiles* are points that divide the range of the observations in a sample variable into subranges (in a sequence) with equal probabilities. The most common quantiles are the 4-quantiles (are also called *quartiles*), the 10-quantiles (are also called *deciles*), and the 100-quantiles (are also called *percentiles*).   □

Example 1.23 (Pareto Chart of a Variable) A *Pareto chart* is a bar graph. Figure 1.2a shows a Pareto chart of the three species in the Iris data sample: Virginica, Versicolor, and Setosa. The lengths of the bars are proportional to the frequency of each category and the curve presents their cumulative distribution.   □

Example 1.24 (Histogram of a Continuous Variable) A *histogram* is an approximate representation of the distribution of numerical data. A histogram of the sepal length for the Iris dataset is shown in Fig. 1.3.   □

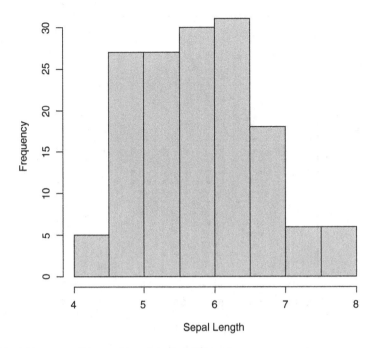

Fig. 1.3 A histogram of the sepal length in the Iris dataset

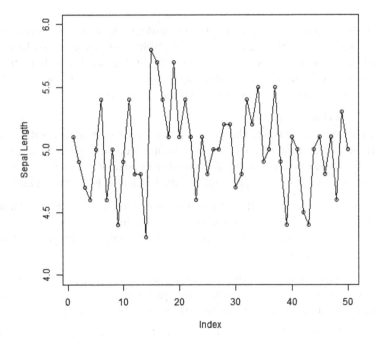

Fig. 1.4 A run chart of the sepal length in the Iris dataset

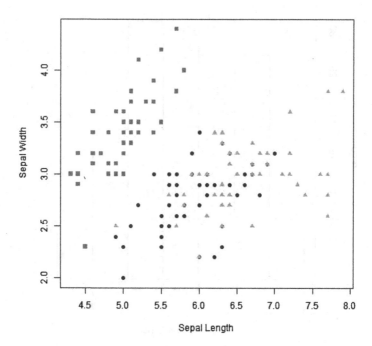

Fig. 1.5 A scatter plot of sepal length and width for the Iris dataset

Example 1.25 (Stem-and-Leaf Plot for a Quantitative Variable) A *stem-and-leaf plot* is a way of presenting a quantitative variable. The plot can also be treated as a table. For example, a stem-and-leaf plot of the sepal length in the Iris dataset is shown in Table 1.3. By studying a stem-and-leaf plot, one can achieve a visualization similar to the one obtained from a histogram.   □

Example 1.26 (Run Chart)
   To understand the dynamic change of a variable over time, a run chart is usually used. If the first 50 observations are recorded in a sequence over a time period of time for the Iris dataset, one may study whether the size of the flower will change over time. A run chart of the sepal length for the Iris dataset is shown in Fig. 1.4.   □

Example 1.27 (Scatter Plot) It is common to have more than one variable (feature) in a dataset. A *scatter plot* represents an observation by a dot in the figure for two features (*x*-axis and *y*-axis). A scatter plot of the sepal length and sepal width in the Iris dataset is shown in Fig. 1.5.   □

Example 1.28 (Multivariate Chart) A *multivariate chart* is used to combine different visualizations of multiple variables into one single chart to enable an overall comparison. A multivariate chart for the Iris dataset is shown in Fig. 1.6.   □

Example 1.29 (Targeted Projection Pursuit) In big data, there are too many variables in the data corpus. It is hard to visualize all of them at once. *Targeted*

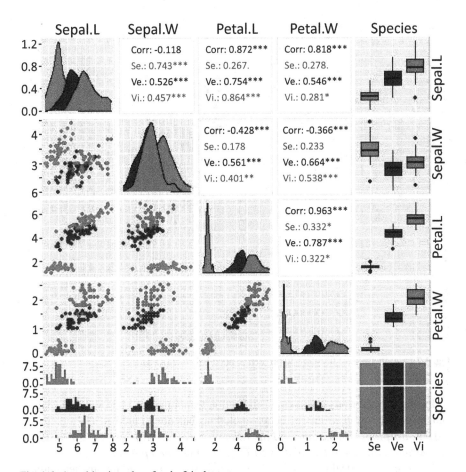

Fig. 1.6 A multivariate chart for the Iris dataset

projection pursuit projects the dataset into a lower dimensional space in order to get a clearer visualization. For example, if one is only interested in the combinations of the four Iris features, say

SepalLength $= 0.4$, SepalWidth $= 0.1$, PetalLength $= 0.9$, PetalWidth $= 0.4$;

and

SepalLength $= 0.7$, SepalWidth $= 0.7$, PetalLength $= 0.2$, PetalWidth $= 0.1$,

one can look at a scatter plot of the two targeted projection pursuit, shown in Fig. 1.7. It is clear that this might be a better visualization for a human to understand the difference/separation between the two species than Fig. 1.5. □

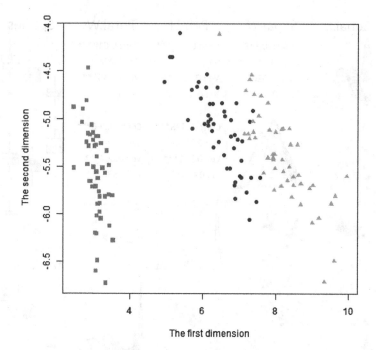

Fig. 1.7 A scatter plot of the targeted projection pursuit for the Iris dataset

In summary, various graphical visualization techniques can be used to visualize and help understand a dataset better, including but not limited to (in alphabetic order): *Boxplots, Histograms, Multivariate charts, Pareto charts, Run charts, Scatter plots, Stem-and-leaf plots*, and *Targeted projection pursuits*.

However, when dealing with big data, even these simple summarizing techniques might not perform well enough in providing a comprehensive picture. Other techniques, such as clustering and granularity re-scaling, may be useful, but they are applications of analytic techniques subject of the coming chapters, so they will be discussed in Chap. 9.

1.2.5 Data Understanding

Another way to understand data is through the use of analogies and abstractions, similar to the way life is understood in the form of specific organisms (my dog or cat), then abstractions into groups (like the biological class of all domestic cats or dogs), then even more abstract concepts (like mammals). Abstractions that are particularly pervasive and powerful among humans involve the concepts of *space* and *time*. They are fundamental to humans and much of intelligence has to do with that (even to the point that some influential philosophers like Kant have proposed that any concept has to be ultimately translated into such geometric concepts to really make sense to the human mind ([1] has more details). So, it is not surprising that mathematicians, physicists, and even cognitive scientists have developed conceptual tools to understand them. They are particularly useful in data science, so it is worthwhile to take a closer look.

The starting point is the mathematical notion of Cartesian coordinate systems familiar to most high school students. They are the usual Euclidean spaces used in calculus and analytic geometry. Geometry is about points, lines, planes, distances, and relationships between objects in terms of them (e.g., shape). (Although the so-called synthetic geometry does not make use of metric rulers to measure distance, as is the case of classical Euclidean geometry from high school, only analytic geometry and its generalizations will be of concern here.) Psychologists have discovered that humans use the concept of distance very loosely. For example, in a city, the distance to go from point A to point B is not the length of the straight line segment joining them because one cannot walk through buildings but has to follow the rectangular street layout of the city. Likewise, if one is flying across the continent between New York and San Francisco, the shortest straight route would require to dig a tunnel through the round Earth, so a pilot will rather take a course along a segment of a greatest circle in the air parallel to the surface of the Earth. Hence, mathematicians have concluded that a more precise concept is required; one that can be characterized as an *abstract concept of distance* in a given space $\mathbf{E}$ as follows.

A *distance* (or *metric*) is a function

$$d(*, *) : \mathbf{E} \times \mathbf{E} \longrightarrow \mathbf{R}$$

assigning a nonnegative real-number $d(x, y)$ to every pair of points $x, y \in \mathbf{E}$ so that three properties are satisfied for arbitrary points x, y, z:

- *Reflexivity*: $d(x, y) = 0$ if and only if $x = y$
  (different points must be at positive distance, not 0);
- *Symmetry*: $d(x, y) = d(y, x)$
  (distance is adirectional);
- *Triangle Inequality*: $d(x, z) \leq d(x, y) + d(y, z)$
  (it is always farther to make a stop at y on the way from x to z).

A set of points $\mathbf{E}$ endowed with a distance function is called a *metric space*. (To ease notation, $d(x, y)$ may be just denoted $|x, y|$ in the sequel.)

Mathematicians feel quite comfortable with it because they can reason about any such notion just like they would about ordinary distance in Cartesian geometry, as long as the three key properties above are satisfied for arbitrary points $x, y, z \in \mathbf{E}$. Likewise for ordinary people and pilots moving around cities on Earth.

Therefore, the city distance in a Euclidean plane between two points is fine, and different metrics in the Euclidean space can be used to endow it with a geometric structure for a variety of purposes. Real datasets that could be mapped into a metric space can be analyzed by leveraging its structural properties that would otherwise be invisible to us. This kind of approach will prove invaluable for dimensionality reduction in upcoming chapters.

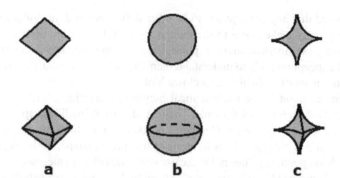

Fig. 1.8 Balls (sets of points within a given radius r from a center **c**) in ℓ_p spaces for (**a**) $p = 1$; (**b**) $p = 2$; (**c**) $p = 3$ in the Euclidean 2D plane (circles, *top*) and 3D space (spheres, *bottom*). The balls can be closed $B[\mathbf{c}, r]$ or open $B(\mathbf{c}, r)$, depending on whether they include the points at distance exactly r (on their boundary) or not, respectively

Example 1.30 In a Cartesian 2D plane, the distance between two points $x = (x_1, x_2)$ and $y = (y_1, y_2)$ given by

$$\ell_1 : |x, y| = |x_1 - y_1| + |x_2 - y_2|$$

is called the *Manhattan distance* and is the ideal definition of a city distance where unit squares stand in for city blocks. The reader may notice that this distance has been obtained by substituting the exponent $p = 2$ in the ordinary Euclidean distance in Cartesian geometry by $p = 1$. The three properties of a metric still hold, although we need to adjust our intuition a little. For example, in the ordinary distance, the disk of radius 1 centered at the origin (i.e., the set of points within distance 1 from the origin) looks like an ordinary circle, whereas in the Manhattan distance, the same concept turns into a diamond since an equation $|(x, y), \mathbf{0}| = |x| + |y| = 1$ defines line segments joining the four points $(\pm 1, \pm 1|$, as shown in Fig. 1.8.   □

Example 1.31 The Manhattan distance is sometimes called the ℓ_1 distance because it is in fact just one in a family of distance functions $\{\ell_p\}_p$ indexed by positive integers $p > 0$ and given by

$$\ell_p : |x, y|_p = [\, \Sigma_{1 \leq i \leq n} \, |x_i - y_i|^p \,]^{1/p}.$$

The particular case $p = 2$ is actually identical to the ordinary Euclidean distance in Cartesian geometry (the square root of the sum of squares of the coordinate differences) and $p = 1$ gives the Manhattan distance.   □

These abstractions enable familiar concepts in a human mind to understand and reason other apparently unrelated concepts (like the concept of a circle and a diamond being abstractly as similar as cats and dogs are both mammals), as will become more and more evident in the following chapters.

1.3 Populations and Data Sampling

This section describes how to cope with the fact that populations are usually large and inaccessible in the whole. Even if they became available, computational constraints to find solutions usually prevent full use of them anyway, despite the fact that what is needed are models that *answer questions for arbitrary and unknown elements of the population!*

Solutions to problems require either (rarely met) assumptions on the target population or lots of data to train models that may help answer the questions. Since collecting a full or even a significant sample (about a subset) from the population is commonly materially infeasible, sampling techniques are crucial to collect unbiased and representative data. A good sample will contain enough information to represent the key properties/characteristics of the population necessary to solve the problem for the entire population (not just the sample). This is easier said than done because

- Data collection is subject to available data and usually requires measurements with available technology;
- Data acquisition is very costly in terms of effort and time;
- Data is ephemeral and constantly requires updates (even for historical data as a result of new research!).

Therefore, sampling a population for good data requires careful planning and carefully chosen methods.

1.3.1 Sampling

Example 1.32 (Political Poll/Economic Survey) One may want to make a prediction about the result of the next presidential election. A single opinion is hardly a good prediction, but obviously, it is impossible to survey all individuals or make a proper and random selection from all citizens of the USA. With a proper sampling, a total sample size of 1500–2000 would be sufficient to achieve, say 95% confidence about the prediction of the outcome of the election in a population of over 200M eligible voters. □

Example 1.33 (Biomedical Study) Methylation of cytosine residues of cytosine-phosphate-guanine dinucleotides (CpGs) is one of the important DNA-based biomarkers associated with aging and/or some diseases. For any individual, there are many CpG sites (out of 28,000,000 CpGs with over 680,000 "useable" CpGs) that can be collected with all the CpGs values normalized to the range from 0 to 1. Because of the high cost of the data collection, only a few thousands of individuals can be sampled. The data can be used to build an "epigenetic clock" to predict epigenetic age, or to find key biomarkers associated with some specific diseases. □

For these examples, a "good" random subset (sample) should include similar characteristic of key variables (e.g., gender, race, age, socio-economic status). However, there are major differences between these two examples in terms of the goals of the study. The first example is typical in *survey sampling*, where the main interest is to estimate the population characteristic (e.g., election result), whereas the second example is typical in data science, where the major goal is to build a predictive model between response (e.g., epigenetic age or specific disease status) and the input features (e.g., CpG sites). In particular, the major goal of this book is to describe techniques to reduce the number of variables (factors or features) necessary to build a predictive model that yields results comparable with or even better than a model obtained from many more features.

In general, feeding low quality data to a solution will produce, unsurprisingly, a low quality solution (also known as *Garbage in? Garbage out!*). A good sample should exhibit four characteristic properties (R^2IBS):

- *Relevant*
- *Representative*
- *Informative*
- *Barely Sufficient*

Traditional sampling techniques were guided by statistical methods and usually assumed that the population is finite. Sampling techniques can be classified according to several criteria. One of them is whether a probabilistic model is considered in choosing the datapoints. Probabilistic sampling selects elements assuming a uniform distribution, e.g., every element of the population gets an equal chance to be selected in the sample. Others include the types of sampling summarized in Table 1.4

Usually, a combination of stratified sampling or cluster sampling and simple random sampling is used. The key advantage of probabilistic sampling is that it usually produces good results, e.g., meaningful statistical inferences on the population based on the sample collected.

Table 1.4 Probabilistic sampling methods for data acquisition

Sampling	How
Simple random sampling	Select points randomly assuming a uniform distribution.
Stratified sampling	Divide the population into strata (e.g., gender, regions) and do simple sampling in each stratum.
Systematic sampling	Index the population, select a first datapoint and then select every kth datapoint after until a certain sample size is achieved.
Cluster sampling	Divide the population into a few groups (e.g., family, genus), then sample each group.
Multistage sampling	If the population cannot be indexed, divide the population into stages (e.g., states or counties), then sample each stage.

Table 1.5 Nonprobabilistic sampling methods

Sampling	How
Convenience sampling	Select datapoints that happen to be most accessible to the researcher.
Purposive (judgement) sampling	Select datapoints that appear most useful for the problem.
Quota sampling	Nonprobabilistic version of stratified sampling.

When no probability distribution is available for the population, the sampling techniques include those in Table 1.5. However, convenience sampling is unlikely to yield a representative sample, so it cannot produce generalizable results.

1.3.2   Training, Testing, and Validation

Most of the data collected in the field of data science are ad-hoc "observational" and so lack proper consideration of the methods used for sampling. Thus data collection is usually done by nonprobabilistic sampling. As pointed out above, the main goal of data science is to build a good predictive model for the response/target variable based on input variables/features in the data. On the other hand, in survey applications, other random sampling schemes play a major role for inference/estimation of population characteristics (e.g., mean or proportion) of interest. Therefore, in statistical applications, it is common to treat the observed big data as the "population" of interest and use probabilistic sampling techniques to select a so-called *training sample* from this "population" to build a model and evaluate the model accuracy on another selected *testing sample*. In this case, one can apply various random sampling strategies (e.g., stratified random sampling) to choose "good" training and testing sets for representative samples. This is a key element to build a better model without the problem of over/under model fitting. Furthermore, one can also choose yet a third dataset for the purpose of "validation" which is commonly used for the purpose of model selection to choose from among competing candidate model solutions. Such a sample is called a *validation sample*.

Since random sampling schemes are used in choosing various samples (training, testing, and validation), the model built and its performance are expected to exhibit random variations/fluctuations, especially for data of small/moderate sample sizes. To obtain a more reliable performance measure, the entire procedure can be repeated a few times to combine these results by taking simple averages or using other advanced ensemble methods. With the decreasing computational cost and increasing power of parallel processing, this may become standard practice in the future.

Finally, random sampling is also used in *cross-validation* (CV) methods used to confirm a model's performance with random multiple "folds" (subsamples) of the training data against the remaining data points in the sample. This is clearly different from the validation sample because it is separate (probably disjoint) from the training sample. CV has its origin in the *Leave-one-out* procedure used to build and test a model leaving out one observation at a time (clearly infeasible when the data size is moderately large or huge). *Stratified cross-validation* (SCV) is another variation of CV where the data is split into folds, with some stratification, on several subpopulations (e.g., gender or some rare event cases), so that each fold is representative of the whole dataset. Using stratified random sampling for some complex datasets can help to maintain the same proportion of different classes in each fold.

On the other hand, in machine learning and data science, as pointed out before, models are expected to have predictive power independently of the sample data used to obtain them. The population is then assumed to be unknown and possibly infinite and a sample is selected once and for all, then split into disjoint subsets as needed (including training, testing, and/or cross-validation). (Sect. 1.4 has further details.)

1.4 Overview and Scope

The overarching goal of this book is to provide a practical and fairly complete, but not encyclopedic, review of Data Science (DS) through the lens of Dimensionality reduction (DR). It approaches data science from the standpoint of applications and problem solving, while also providing prerequisite unifying theoretical foundations and case studies. We have strived to capture the greatest value for professionals seeking to solve problems in their domains of interest, including representative sample datasets and readily available tools to produce tested solutions of high quality.

The intended target audience consists of professionals in any domain science where data science can help in solving problems and answering questions. Domain science is a fairly vague technical term that refers to a specialized area of human knowledge (the domain) characterized by specific questions about a certain aspect of reality (like *what is motion* in physics, *what are physical objects made of* in chemistry, *what is life* in biology, and so forth). In addition to the well-established sciences (physics, chemistry, biology, and their subdomains), they include just about any area where data can be recorded and analyzed to answer questions concerning the individuals or objects the data is about.

Data science presents a singular approach to problem solving when compared to the more established sciences. Traditional sciences are motivated by pressing problems for people to survive and thrive in the world. That requires a degree of understanding of the phenomena involved that enables predictive power. With the advent of computer science and its abstractions into the information age (as embodied by the internet and web, for example), tools were created that can be used regardless of the specific domain. Once this threshold is crossed, then it is natural to group and develop methods and platforms to do this kind of generic science, hence data science. It is a natural next step from mathematics and statistics to *synergistically combine them with the powerful computational tools developed*

in computer science to create a new science that is more than the sum of the parts.
The basic background concepts required from these building blocks are summarized
in Chap. 11. We have strived to leave our niche hats at the door and present a
integrative and synergistic approach that captures the best of the three worlds. That
is the pervasive thread that readers will discover through examples and methods
throughout the book.

1.4.1   Prerequisites and Layout

The content of the book is presented using the same template approach in every
section. Sections begin with an intuitive example of a problem to be solved by
the concepts being introduced in that section. A professional with an undergraduate degree in any quantitative science should be able to follow this part. We
have assumed that the reader is familiar with basic undergraduate mathematics
(multivariable calculus and linear algebra), including matrix algebra. Likewise, we
have assumed that the reader is familiar with the basic concepts in statistics and
probability, including sample spaces, probability distributions, random variables,
and the main results associated with them. Nevertheless, a refresher summary
is given in the Appendices in Sects. 11.1 and 11.2. Since readers may be less
familiar with basic computational background, a summary is likewise made of basic
concepts in computer science as well in Sects. 11.3 and 11.4

These motivating examples in a section are then followed by precise definitions
of the technical concepts and presentation of the results in general situations. That
requires a degree of abstraction that can be followed by re-interpreting the general
terms like in the original example(s). Finally, each section closes with solutions
to the original problem afforded by these techniques, perhaps in various ways to
compare and contrast advantages and disadvantages of the various DR techniques
based on quantitative and qualitative assessments of the solution(s) in the real world.

1.4.2   Data Science Methodology

Solving a data science problem typically requires several steps:

1. *Define the problem precisely.*
   A good definition requires a clear distinction between the *WHAT* are one is trying
   to do, versus *HOW* one is going to actually find a solution. The definition should
   first make sense in the real world to common people who know nothing about DS
   but have a *goal* to achieve. One can then say this definition is more like a *business*
   definition. It should answer fundamental leading questions such as: *WHAT needs
   to be changed? WHAT is the desired outcome?* Figure 1.9 illustrates the point.
   An analogy to taking a trip is most appropriate. It is common to hear people say,

Fig. 1.9 (a) Defining a data science problem appears easy, but (b) is a difficult task, because it has to be distinguished from HOW to solve the problem. WHAT goal to achieve is a destination that has to be decided first because it determines practically everything else and makes decisions easier

let us just get the data first, then we will see what we can do. That amounts to saying, let us get in the car and start driving, we will decide on a *destination* later. A moment's reflection will make it obvious that this is nonsense really. Knowing where one wants to go will lead her some place with some nonzero probability (even if it is the wrong place!). Not knowing where to go should be expected to prove disastrous. An objectively defined and clear *destination* is paramount, possibly including a criterion of quality for a solution to be good enough and acceptable.

Examples of proper ways to define a problem have been given in earlier chapters and more examples are shown in Sect. 11.4. The easiest way is to identify the problem as one of the standard problems in Data Science (Classification? Prediction? Clustering?) or reformulate the problem into a related problem of this kind that might help.

2. *Name a destination.*

Naming a destination determines virtually everything else (HOW to get there, what to pack to wear, and so forth). In Data Science, it will help decide what kind of data needs to be gathered or obtained, what kind of solutions could be tried. There are no preset recipes (contrary to traditional sciences), just a set of tools to try. What is amazing is that this approach usually lands one with some good enough (although perhaps not perfect) solution.

3. *Select an evaluation metric.*

As lord Thompson (the inventor of the Celsius temperature scale) said, one cannot really know something unless one can quantify it. The next step is how to decide if a potential solution is good or not. A clear definition of the problem will make it fairly obvious (they are described further below in this section). Since the problem definition spelling a business goal to achieve was clearly defined in the first step, the definition should provide a criterion whether the solution is good

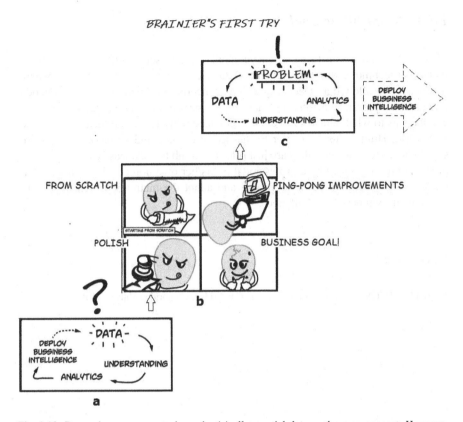

Fig. 1.10 Data science appears to have the (**a**) all-essential data as the core concept. However, upon reflection, (**b**) it is really about problem identification and problem solving because (**c**) the problem being solved dictates not only what and how much data is appropriate but also whether we have gone enough around the loop to be able to deploy a solution for it that is viable in the real world

enough or not (not whether it is optimal). If not, one may need to try to go through this loop again, as shown in Fig. 1.10.

4. *Business Intelligence*

   Finally, an acceptable solution requires re-insertion and implementation back in the real world. That will be the real test whether the solution really makes sense, i.e., whether the problem has been solved or not. If not, one needs to go through the process all over again, with appropriate refinements for a better chance.

1.4.3   Scope of the Book

This book is not meant to be an encyclopedia of data science that includes every method or technique known to humans. Data science is too young to even tell where it is headed to allow that. Deeper questions concerning related matters are touched upon in an exploratory manner in Chap. 10, but only to give the reader a take-home epiphany as to what data science and/or dimensionality reduction mean back in the real world. Humans usually refer to that as *experience* and *knowledge*. They are simply heuristics, because they are just rules of thumb that can easily fail with some practical problem faced by a professional in his/her own domain. With this caveat, readers can use it as a guide to choose and adapt the methods presented here to tackle their own personal challenges.

Reference

1. Broad, C. D. (1978). *Kant: An introduction*. Cambridge: Cambridge University Press.

Chapter 2
Solutions to Data Science Problems

Deepak Venugopal, Lih-Yuan Deng (iD)**, and Max Garzon** (iD)

Abstract This chapter presents a review of statistical and machine learning models to tackle data science problems, arguably the most popular approaches. Both supervised and unsupervised algorithms are described along with practical considerations when using these methods. Empirical results on exemplar datasets are also presented where applicable to illustrate the application of these methods to real-world problems.

2.1 Conventional Statistical Solutions

To study whether a feature can help in solving DS problems, most statistical models formulate the problems into probabilities of mass/density functions in various perspectives. By modeling the probabilities of getting the situations of interest, practitioners can select important features, describe the relationship among variables, make an inference, and classify, predict, or cluster future events.

This section presents a summary of various classical statistical solutions that can be used to solve a problem by building a model or reduce the dimension of the feature space.

2.1.1 Linear Multiple Regression Model: Continuous Response

A number of classical statistical methods can be used in many cases to provide a better foundation for finding solution models and improve their predictions.

D. Venugopal (✉) · M. Garzon
Computer Science, The University of Memphis, Memphis, TN, USA
e-mail: dvngopal@memphis.edu; mgarzon@memphis.edu

L.-Y. Deng
Mathematical Sciences, The University of Memphis, Memphis, TN, USA
e-mail: lihdeng@memphis.edu

29

In the notation of Chap. 1, given n data points $(\mathbf{x}_1, y_1), \ldots, (\mathbf{x}_n, y_n)$, where $\mathbf{x}_i = (x_{i1}, \ldots, x_{ip})$ is a covariate vector for the ith observation of predictor features and y_i is the target response, it is common to represent the dataset in matrix form

$$\mathbf{X} = \begin{bmatrix} x_{11} & \ldots & x_{1p} \\ x_{21} & \ldots & x_{2p} \\ & \cdot & \\ & \cdot & \\ & \cdot & \\ x_{n1} & \ldots & x_{np} \end{bmatrix} = [X_1 \, X_2 \, \cdots \, X_p] \quad \mathbf{Y} = \begin{bmatrix} y_1 \\ y_2 \\ \cdot \\ \cdot \\ \cdot \\ y_n \end{bmatrix},$$

where the covariate matrix can be viewed as n row vectors $(\mathbf{x}_1, \mathbf{x}_2, \ldots, \mathbf{x}_n)$ of dimension p, or p column vectors of dimension n $(X_1, X_2, \ldots, X_p)$ and $\mathbf{Y}$ is the response column vector of dimension n, $(y_1, y_2, \ldots, y_n)'$.

If the response column vector $\mathbf{Y}$ is continuous, it is common to assume at first that a solution model is given by

$$\mathbf{Y} = f(\mathbf{X}) + \boldsymbol{\epsilon} = f(X_1, X_2, \ldots, X_p) + \boldsymbol{\epsilon},$$

where $f(\cdot)$ is a function to be estimated and $\boldsymbol{\epsilon}$ is a random variable (Sect. 11.1 gives some probabilistic and statistical background) for the error in the estimation. Clearly, the general linear model is a special case with a linear function $f(\cdot)$

$$\mathbf{Y} = \sum_{j=0}^{p} \beta_j X_j + \boldsymbol{\epsilon}$$

and $X_0 = 1$.

When the dimensionality of the data (the number of columns p of $\mathbf{X}$) is large, one may consider some variable selection procedure to find a reduced model with fewer significant variables, or one may use leading principal dimensions from PCA (principal component analysis, described in Sect. 4.1). Popular *feature selection* and *subset selection* methods are:

1. Best subset
2. Stepwise selection: forward and backwards

A common selection criterion for assessing quality of fit of a model is the classical R^2 or the root mean-squared error (RMSE), given by

$$R^2 = 1 - \frac{\text{SSE}}{\text{SST}} = 1 - \frac{\sum_{i=1}^{n}(y_i - \hat{y}_i)^2}{\sum_{i=1}^{n}(y_i - \bar{y})^2}$$

and

$$\text{RMSE} = \sqrt{\text{MSE}}, \quad \text{MSE} = \frac{\text{SSE}}{n - p}.$$

Clearly, R^2 or RMSE for assessing quality of fit is very suitable because they tend to choose models with more predictors.

Hence, a better quality criterion is needed that takes into account the size of the model since small models are preferred that still fit well, even if one has to sacrifice a small amount of "goodness-of-fit" for a smaller model. Three more criteria are available, namely AIC, BIC, and Adjusted R^2.

2.1.1.1 Akaike Information Criterion (AIC)

Another measure of the quality of a model is the AIC, defined as

$$AIC = n \log \left(\frac{RSS}{n} \right) + 2p,$$

where RSS $= \sum_{i=1}^{n} (y_i - \hat{y}_i)^2$ and $\hat{\beta}$ and $\hat{\sigma}^2$ and p is the number of β parameters in the model. The term $2p$ is called a penalty component of the AIC because it is large when p is large, while the aim is to find a small AIC (small RSSD and p).

Thus, a good model will strike a good balance between the conflicting goals of fitting well and using a small number of parameters. The smaller the AIC, the better.

2.1.1.2 Bayesian Information Criterion (BIC)

The Bayesian Information Criterion BIC is similar to the AIC but has a larger penalty:

$$BIC = n \log \left(\frac{RSS}{n} \right) + \log(n)p.$$

BIC also quantifies the trade-off between a model that fits well and the number of model parameters, although, for a reasonably large sample size, it generally picks a smaller model than AIC. As with AIC, the model with the smallest BIC is selected.

The penalty for BIC is $\log(n)p$ rather than the AIC's penalty of $2p$. Therefore, for any dataset where $\log(n) > 2$, the BIC penalty will be larger than the AIC penalty, and thus BIC will be likely to lead to a smaller model.

2.1.1.3 Adjusted R-Squared

It is common to make an adjustment to the popular R^2 as

$$R_a^2 = 1 - \frac{SSE/(n - p)}{SST/(n - 1)} = 1 - \left(\frac{n - 1}{n - p} \right) (1 - R^2).$$

Unlike R^2 that can never become smaller with added predictors, this adjusted R^2 effectively penalizes for additional predictors and can decrease with added predictors. As with R^2, larger is still better for the adjusted R^2.

2.1.2 Logistic Regression: Categorical Response

If the response column vector $\mathbf{Y}$ is binary vector of 0s and 1s, it is common to consider a generalized linear model of the form

$$g(E(\mathbf{Y}|\mathbf{X})) = g(\mu) = f(\mathbf{X}) + \epsilon = f(X_1, X_2, \ldots, X_p) + \epsilon,$$

where $g(\cdot)$ is a link function to be chosen and $f(\cdot)$ is a response function. Several common choices can be considered:

1. **Logistic regression** (LR)

   It is a very popular model for binary response with the logit link function, $g(\mu) = \ln[\mu/(1-\mu)]$, and

   $$g(\mu) = \ln\left[\frac{\mu}{1-\mu}\right] = \sum_{j=0}^{p} \beta_j X_j + \epsilon.$$

2. **Generalized additive model** (GAM) [16]

   The GAM can be considered an extension of LR without assuming $f(\cdot)$ is a linear response function since it replaces $\beta_j X_j$ with a general smooth function $s_j(X_j)$

   $$g(\mu) = \sum_{j=0}^{p} s_j(X_j) + \epsilon.$$

Typically, GAM is generally used for nonlinear regression problems with continuous response, but it can also be used to build a binary response classifier. Hastie [16] first proposed a generalized additive model (GAM). The GAM can be adapted to different situations, as generalized linear regression can be used when different link functions are utilized. While there is no limit on the choice of the link function $g(\mu)$, the "logit" link is commonly used for a binary classifier.

2.1.3 Variable Selection and Model Building

High-dimensional statistical problems are quite common in various fields of science, and variable selection is important in statistical learning and scientific discovery.

The standard procedure of best subset selection methods is based on AIC or BIC as variable selection criteria are suitable only for a moderate number of input variables. For high-dimensional data, these procedures can be very computationally expensive. Procedures using penalized likelihood methods have been successfully developed recently to deal with high-dimensional datasets. In addition to variable selection, these methods can also be used in estimating their effect in high-dimensional statistical inference. (Further discussion is given in Chap. 8.)

Model building to solve a problem is a process of finding the appropriate relationship between response and input variables. Such a relationship could be either a simple linear relationship or a complicated nonlinear relationship. In addition to the first-order linear model, it is possible to consider other models such as polynomial regression models or generalized additive models (GAMs). One of the major problems for building models for high-dimensional data is the problem of multi-collinearity among many input variables. Furthermore, building GAM could be infeasible for a large number of highly correlated input variables. In such cases, one can consider dimension reduction techniques such as principal component analysis (PCA), to be discussed in Chap. 4.

2.1.4 Generalized Linear Model (GLM)

A popular model that includes both continuous and discrete models is the generalized linear model (GLM). The popular logistic regression model is a special case. There are three components in the GLM, namely, a random component, a systematic component, and a link function component.

Random Component
The random component of a GLM characterizes the distribution of the response variable $y_i, i = 1, 2, \ldots, n$ with a general form of the exponential family of distributions (described in Sect. 11.1 for probability background). It is given by

$$f(y_i; \theta_i) = a(\theta_i) b(y_i) \exp[y_i Q(\theta_i)],$$

where $a(\cdot)$ and $Q(\cdot)$ are functions of θ_i and $b(\cdot)$ is a function of y_i.

Systematic Component
The systematic component of a GLM specifies a linear relationship between a transformed parameter (via a link function below) and its input variables. Specifically, for each y_i and p predictor variables $x_{ij}, j = 1, \ldots, p$, the parameter η_i is a linear function of x_{ij} of the form

$$\eta_i = \sum_j \beta_j x_{ij}, \quad i = 1, \ldots, n.$$

Link Function Component

The link function component of a GLM specifies the transformation, say $g\cdot$ (monotone, differentiable), on the mean response of Y_i, $\mu_i = E(Y_i)$, so that

$$\eta_i = g(\mu_i),$$

that is,

$$g(\mu_i) = \sum_j \beta_j x_{ij}, \quad i = 1, \ldots, n.$$

Special cases of the link function are illustrated next.

Example 2.1 (Linear Model/Regression Model) The usual linear model/regression model is included in the GLM formulation with

$$y_i = \sum_j \beta_j x_{ij} + \epsilon_i,$$

where ϵ_i's are i.i.d. $\epsilon_i \sim N(0, \sigma^2)$. The link function $g(\mu) = \mu$ could be the identity link function.

In particular, the canonical link is a link function that transforms the mean μ to the natural parameter

$$g(\mu_i) = Q(\theta_i) = \sum_j \beta_j x_{ij}.$$

□

Example 2.2 (Logit Models for Binary Data) For binary response data $Y \sim B(1, P)$, its probability density function (pdf) can be written as

$$f(y; P) = P^y (1 - P)^{1-y} = (1 - P) \exp\left(y \log \frac{P}{1 - P}\right),$$

where:

- $a(P) = 1 - P, b(y) = 1, Q(P) = \log \frac{P}{1-P}$.
- The natural parameter is $\log \frac{P}{1-P}$ log odds or logit of P.
- GLM is also called the *logit model*.

□

Example 2.3 (Poisson Loglinear Model) When the response Y represents data counts, it is common to assume a Poisson distribution (described in Sect. 11.1) in the Poisson Loglinear Model $Y \sim Poisson(\mu)$, with probability distribution given by

$$f(y; \mu) = \frac{e^{-\mu} \mu^y}{y!} = e^{-\mu} (1/y!) \exp(y \log \mu), y = 0, 1, 2, \ldots,$$

Table 2.1 Generalized
Linear Models (GLMs)

Random	Link	Systematic	Model
Normal	Identity	Continuous	Regression
Binomial	Logit	Mixed	Logistic regression
Poisson	Log	Mixed	Loglinear

Table 2.2 Dataset with three
features to predict whether a
person likes a restaurant or
not

Price	Fast	On-campus	Likes
High	Yes	Yes	No
Low	Yes	Yes	Yes
High	Yes	No	No
High	No	Yes	No
Low	Yes	No	Yes

where:

- $a(\mu) = e^{-\mu}, b(y) = (1/y!), Q(\mu) = \log \mu$.
- The natural parameter is $\log \mu$, and the canonical link function is $\eta = \log \mu$.

$$\log \mu_i = \sum_j \beta_j x_{ij}, \quad i = 1, \ldots, n.$$

□

A summary of the various kinds of GLMs is shown in Table 2.1.

2.1.5 Decision Trees

Decision trees are models that use a tree structure to solve a classification problem. Specifically, the internal nodes in the tree represent questions about the values of features in the data, and the leaf nodes represent classes. Each feature is branched (or conditioned) on different possible values for that feature. Given a dataset as in Table 2.2, an example decision tree is shown in Fig. 2.1. To classify a data instance, starting with the root node, each node is a decision point where one of the branches is selected based on the feature value in that instance. The leaf nodes in the decision tree (nodes with no children) correspond to class labels. For instance, for the example tree shown in Fig. 2.1, to classify whether a patron likes a restaurant or not, if for a specific data instance, Price = "high," Fast = "no," and On-campus = "no," the decision tree will make decisions to follow branches corresponding to the feature values from root to leaf and output a class Likes = "No".

Decision trees are highly versatile models since they can implement any Boolean function. Specifically, given the truth table for any Boolean function, each path in the decision tree can encode one row in the truth table of the Boolean function. However, the size of a decision tree can be large when learning a complex classifier. That is, a large number of nodes may be required to express the classifier function.

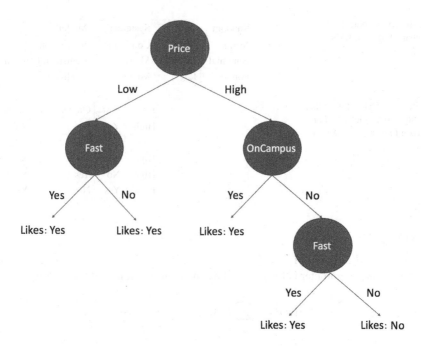

Fig. 2.1 A decision tree (DT) classifier for the restaurant dataset in Table 2.2

The main goal in learning a decision tree from data is to learn a *compact* tree that has as few nodes as possible. Several heuristics used to construct such trees are described next.

Decision tree learning (sometimes called *induction*) incrementally grows the decision tree given a training dataset. Typically, a greedy approach is used for tree induction, where in each step, a feature is selected as a node in the decision tree. The selected feature essentially splits the data into different branches based on the possible values that the feature can take on. Therefore, in each step, the main computational task is to select a "good" feature, i.e., a feature that acts as the best classifier for the data. For example, among the three features in for dataset given in Table 2.2, if the feature Price always determines whether one likes a restaurant or not, then this feature encodes the most useful information required for classifying the data and should be considered ahead of the other features while learning the decision tree. To select the best feature in each step of decision tree learning, a splitting criterion is used to score the utility of each feature.

While there are several different possible heuristics that can be used as a splitting criterion, one of the most widely used decision tree learning algorithms, ID3 [22], and its variants use conditional entropy in the splitting criteria. Entropy is a measure of uncertainty or randomness in a sample, to be defined precisely in Sect. 6.1. For example, given a dataset where all the instances belong to the same class, the entropy of this dataset is 0 since there is no randomness. At the other extreme, consider two possible classes where half of the dataset belongs to one class and the other half belongs to another class; in this case, the entropy is maximum. The conditional entropy for a feature is a measure of randomness with respect to the class label after splitting or conditioning the data according to each value of the feature. Thus, in the example in Table 2.2, the conditional entropy for Price is 0 since for each value of Price, the data samples have the same class label. On the other hand, for the feature On-Campus, for each of its values, the data samples are equally distributed among the two class labels. Therefore, the conditional entropy for the On-Campus feature is large. In other words, if $H(D)$ stands for the entropy of data D given c classes in D, the entropy of D is given by

$$H(D) = \sum_{i=1}^{c} -P_i \, \log_2 P_i \, ,$$

where P_i is the proportion of D belonging to class i. If the splitting/conditioning of D is performed using feature x_i, the conditional entropy after this split is given by

$$H(D \mid x_i) = \sum_{v \in Values(x_i)} \frac{|D_v|}{|D|} H(D_v),$$

where $Values(x_i)$ is the set of possible values for feature x_i, $D_v \subseteq D$ where feature x_i has the value v. Thus, the conditional entropy for x_i is computed by computing the entropy values for subsets of the data specific to each value of x_i and then computing a weighted average of these values. It is easy to see that if a feature splits the data such that for each value of the feature, the class labels are uniform (for example, the feature Price in the aforementioned example), then the conditional entropy is equal to 0. Thus, a feature with smaller conditional entropy can classify the data more effectively than a feature with larger conditional entropy.

The decision tree algorithm proceeds as follows. In each step, the feature with the smallest conditional entropy is selected. The selected feature is then used to partition the data D, into splits $D_1 \ldots D_k$, where each split of the data corresponds to a specific value of that feature. This process is repeated recursively for $D_1 \ldots D_k$.

Some practical considerations of decision tree learning include:

- To compute the conditional entropy, the features are assumed to be discrete. Continuous features are thresholded into discrete values to approximate the conditional entropy.
- Entropy computations are computationally expensive. Therefore, a common problem in decision trees is that scaling them to big data is expensive. A simple solution to this is to compute the conditional entropy values from a smaller sample of the data.
- A significant advantage of decision trees is that they are perhaps among the most interpretable machine learning models. That is, a human user can understand the output of the decision tree, and for each classification that the decision tree

makes, a trace of the sequence of decisions that led to that classification can easily serve as a causal chain of reasoning for the label assignment. Therefore, decision trees are among the leading models in applications where interpretability is important such as healthcare [5] and business analytics [19].

2.1.6 Bayesian Learning

The Bayes theorem is among the most fundamental theorems in probability theory (Sect. 11.1 defines background concepts in probability and statistics and gives a precise statement). It can be applied for classification by assigning the most probable class value given a set of features. Specifically,

$$C' = \arg \max_{C} P(C|\mathbf{x}),$$

where $P(C|\mathbf{x})$ is the probability of instance $\mathbf{x}$ having class label C. Using Bayes theorem, one can rewrite the above results in the following classifier:

$$C' = \arg \max_{C} P(\mathbf{x}|C) \frac{P(C)}{P(\mathbf{x})},$$

where $P(C)$ is the probability of the class C and $P(\mathbf{x}|C)$ is the probability of observing features $\mathbf{x}$ given that the class is C. The denominator on the right-hand side of the above equation is immaterial to compute the maximum, so the classifier can be re-written as

$$C' = \arg \max_{C} P(\mathbf{x}|C) P(C).$$

Given training data, estimating $P(C)$ is quite straightforward. $P(C)$ is estimated by computing the percentage of instances in the data with class value equal to C. Specifically, if the total number of training instances is n and among these, the number of instances with class label C is n_c, then the estimated probability is $P(C)$ = $\frac{n_c}{n}$. On the other hand, estimating $P(\mathbf{x}|C)$ is much harder. This probability is the percentage of instances that have feature values exactly equal to $\mathbf{x}$. To compute this, the data needs to contain sufficient instances of any configuration of feature values, which may not happen in practice. For example, if there are 10 binary features (0/1), there are $2^{10} = 1024$ possible configurations of feature values given a class label, and the training data would need to contain sufficient instances to estimate all 2^{10} probabilities. As the number of features increase, the number of probabilities needed to find the classifier grows exponentially.

To scale up Bayesian learning, the *Naive Bayes* classifier makes a simplifying assumption that the features are conditionally independent given the class. This means that the joint probability (defined in Sect. 11.1) over the features given a class can be expressed as a product of probabilities over each feature as follows:

$$P(C|\mathbf{x}) = \frac{P(\mathbf{x}|C)P(C)}{P(x)} \propto P(\mathbf{x}|C)P(C) \tag{2.1}$$

$$\propto P(x_1, x_2, \ldots, x_K|C)P(C) = \prod_{i=1}^{K} P(x_i|C)P(C),$$

where $P(x_i|C)$ is the probability of a single feature value x_i given class C. In this case, considering the previous example, 10 features only require 10 probabilities corresponding to each class instead of 2^{10} probabilities, an exponential reduction that allows the classifier to scale up to a large number of features.

Learning the Naive Bayes classifier is straightforward where the probability of a feature value given the class is simply the proportion of training examples of the class where the feature takes that specific value. For example, if the classification task is to identify spam/nonspam emails and one of the features is the word "free," where "free" occurs in 10% of the nonspam emails and 50% of the spam emails in our training data, independently of other features (based on an assumption of conditional independence), then $P(\text{"}free\text{"}|Spam) = 0.5$ and $P(\text{"}free\text{"}|NonSpam) = 0.1$. In general, if the total number of training instances of class C is n_c and among these, feature x_i has value v in n_v instances, the estimated conditional probability is given by

$$P(x_i = v|C) = \frac{n_v}{n_c}. \tag{2.2}$$

Some practical considerations in using the Naive Bayes classifier include:

- If a feature x_i is continuous, then $P(x_i = v|C)$ can no longer be estimated by counting the proportion of training examples with $x_i = v$. In this case, a variant of Naive Bayes called Gaussian Naive Bayes is used to estimate the conditional probability. Specifically, each continuous feature has a conditional probability that is assumed to be Gaussian, and the Gaussian parameters are learned from the data.
- A probability estimate of 0 for the conditional probability corresponding to a single feature makes the entire product of conditional probabilities in Eq. (2.1) equal to 0. This happens when a particular feature value may be absent in the training data for a given class. A widely used approach to correct zero probabilities is called the *Laplace correction*. The idea is to assume that the feature value occurred a constant number of times in the training data (even though it may not have). The Laplace corrected value in Eq. (2.2) is equal to $\frac{n_v + r}{n_c + r|x_i|}$, where r is a constant and $|x_i|$ is the number of values feature x_i can take on.
- The Naive Bayes classifier makes the assumption of conditional independence even though this assumption is generally false, i.e., in most cases features are not

conditionally independent given the class. However, the Naive Bayes classifier works quite well and in some cases (such as in text classification), where it achieves highly competitive accuracy compared to much more sophisticated models. (A formal analysis on the reasons for its good classification performance despite its assumptions is provided in [8].)

- The Naive Bayes model can be easily interpreted since the conditional probabilities of features can be compared with each other to understand the relative importance of features in the model.

2.2 Machine Learning Solutions: Supervised

Machine learning (ML) methods are arguably one of the most widely used approaches to solve data science problems. A machine learning model can be viewed as a program that improves itself with experience [20]. The next two sections provide a summary review of most commonly used machine learning algorithms. The descriptions focus on practical aspects of these solutions related to solving data science problems, with appropriate references to other sources for a more comprehensive treatment of the technique(s).

ML algorithms can be roughly classified into two major categories, *supervised* and *unsupervised*. Supervised algorithms require that the data contain a designated target (response) feature with the expected answers to the instances of the problem at hand, in addition to a given set of predictor features in a data point, whereas unsupervised methods just require the predictor feature vectors describing the data point.

This section gives an overview of supervised learning algorithms, and the next is concerned with unsupervised learning methods. To ground the abstract concepts, the classification problem for the problem [**CharRC**] of handwritten character recognition and the MNIST dataset (Sect. 11.3 has more details) is used to illustrate the methods throughout this section.

Example 2.4 The classification problem of handwritten digit recognition [**CharRC**] (described in Sects. 1.1 and 11.4) calls for a category label of digits $0, 1, \ldots, 9$ for a given input image (presumably a handwritten digit), as illustrated in Fig. 2.2. Each image is a 28x28 grayscale image of a single digit from actual handwritten ZIP codes. Due to personal variations in writing, naturally the same digit can appear differently in different instances, and while humans can perceive these variations quite easily, coming up with an automated program to do the same is quite challenging. To improve itself automatically and learn, a machine learning algorithm must have a performance measure to decide how to improve. A good choice here can be the accuracy with which the program identifies the handwritten digits over a dataset. Much like how humans become better at a task with practice, to obtain improvement in performance, some training helps a program gain experience in solving the task. This training is provided to the program through data. The key

Fig. 2.2 A sample of images from the MNIST digits dataset for the problem of handwritten character recognition [**CharRC**]

requirement of learning (as opposed to memorization) is that the program not only performs well on data that it has been trained on, but it should also work on new data it has never seen before. In other words, given an unseen image of a handwritten digit, the program should be able to successfully identify the digit, even though that specific image is new to the program. Thus, the program does not simply memorize the data but learns some deeper patterns in the data to make useful inferences about new data. □

These kinds of algorithms are examples of what is known as *supervised learning*. Specifically, in supervised learning, the algorithm is trained with data where each data point includes a label for a specific discrete class giving the correct answer in the problem. In general, the goal of any supervised learning algorithm is to refine successive possible solutions based on data instances and their labels in such a way that a new data instance that it has previously not seen during training is very likely to elicit a correct classification.

Formally, supervised learning is defined as follows. Given n data points $(\mathbf{x}_1, y_1), \ldots, (\mathbf{x}_n, y_n)$, where $\mathbf{x}_i = (x_{i1}, \ldots, x_{ip})$, y_i are vectors representing features for the ith observation and its corresponding class, the supervised learning algorithm is to return a function $f(\cdot)$ defined on the population for the problem so that $y_i = f(\mathbf{x}_i)$.

Several algorithms for supervised learning are reviewed in the remainder of this section.

2.2.1 k-Nearest Neighbors (kNN)

Example 2.5 The key idea in the k-nearest neighbors classifier is illustrated in Fig. 2.3. If each image in the MNIST dataset can be represented as a single point in a Euclidean space, an algorithm can classify a new image by simply using the majority class among all its neighbors within certain radius in this space. This idea is simple and intuitively appealing, but will it produce good results as a learning algorithm? How can the neighborhood be chosen? □

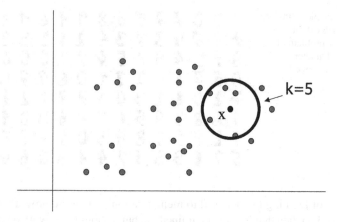

Fig. 2.3 The key idea in the k-nearest neighbors algorithm on the MNIST dataset. Data points belong to two classes (e.g., digits in the MNIST data), and X represents a new point to be classified based on the majority class of $k = 5$ of its nearest neighbors

The **kNN** (k-nearest neighbors) algorithm is a type of *instance-based* learning. In contrast to other solutions such as neural networks or decision trees, instance-based learning methods do not learn a function for classification from training data. Instead, they simply store the training data instances. When asked to classify a new instance, they retrieve similar instances in the training data to help classify this new instance. Thus, the function f is not defined in abstract, but in the context of every instance, making this approach more flexible.

The **kNN** algorithm assumes that each data instance is a point in a pD metric space, where P is the number of features in the data points. Given a new point $\mathbf{x}_j$ to classify, it computes k-nearest neighbors to this point among all the points in the training data and classifies the new instances as the same class as the majority of the k neighbors. Typically, the neighbors are computed using the ordinary Euclidean distance ℓ_2 (defined in Sect. 1.2), although other distance metrics are not uncommon. Naturally, an odd number is chosen as the value of k to guarantee a majority among the class labels in the neighborhood. **kNN** assigns $\mathbf{x}_j$ the same class label as the one that occurs in majority of the k-nearest neighbors in its neighborhood of $\mathbf{x}_k$.

kNN implicitly assumes that all neighbors are equally important. In a variant of k-nearest neighbors, called *distance-weighted* k-nearest neighbors, neighbors to a data point are weighted inversely by their distance to that point, thus giving greater importance to closer neighbors. Further, the Euclidean distance computation is also affected by the scale of each feature. For example, if one of the features is the annual income and the other is credit card rating, naturally the scale of annual income is much larger than credit card rating and dominates the distance computation. Therefore, typically, the features are standardized using Z-scores (defined in Sect. 11.1) before applying the **kNN** classifier (each feature value is subtracted with the mean value for that feature and divided by the standard deviation value for that feature).

Table 2.3 Results for
MNIST varying the
neighborhood in **kNN**

k	Precision	Recall	F1-score
1	0.982	0.982	0.982
3	0.986	0.986	0.986
5	0.993	0.993	0.993

Example 2.6 (k-Nearest Neighbors for MNIST) In the **kNN** solution to the hand-written digits classification problem [**CharRC**], each pixel in the image represents a single dimension in Euclidean space for a **kNN** classifier. Table 2.3 shows the precision, recall, and F1-score (defined precisely below in Sect. 2.4) on a test dataset (chosen to be 25% of the data) for varying k when **kNN** is applied to this problem. As seen here, increasing k can smooth some of the local irregularities in the labels and improves performance.   □

Important practical considerations in applying **kNN** include:

- **kNN** is a computationally expensive classifier since it needs to store/index all the training data. In contrast, classifiers such as decision trees or neural networks learn a model from the training data and do not need to use it again after trained for classification. Thus, although training a classifier is time-consuming, classifying a new instance is easy after training. On the other hand, **kNN** has all the overhead in classifying a every new instance since they need to search over the training data to find the neighborhood of that instance. Finding neighborhoods for a point in higher-dimensional space (when the number of features is large) is a search problem that becomes exponentially harder with the size of the data. Typically, specialized data structures (such as KD trees) are used to quickly find neighborhoods for a data point. Even with these data structures, however, the scalability of **kNN** is quite limited when compared to other methods.
- In most applications, it is generally the case that some features will be more relevant than others. As the number of irrelevant features increases, the distances computed in **kNN** become dominated by these features. This fact is generally termed the curse of dimensionality (discussed further in Sect. 10.3). While most machine learning methods perform poorly in the presence of irrelevant features, **kNN** are sensitive to the curse of dimensionality and typically perform much worse than other approaches with a large number of features.
- Since **kNN** assumes that data points live in a common Euclidean space, all the features in the data must be real-valued. While it is possible to have features that are discrete/categorical, such features need to be embedded in a Euclidean space using an additional step to enable the k-nearest neighbors algorithm.

2.2.2 Ensemble Methods

Ensemble methods refer to a general approach where various classifiers are combined together for better performance.

Example 2.7 A natural approach in medical diagnosis is to use several expert opinions rather than relying on a single doctor's opinion. An illustration of ensemble methods is shown in Fig. 2.4. Although each classifier separately may make a certain degree of error in the classification, the combined classifier that uses the output of all classifiers can correct these errors based on the collective outputs.

Two widely used approaches to combine classifiers include *Bagging* (*Bootstrap Aggregation*) [4] and *Boosting* [25]. These methods try to improve generalization performance of a classifier. Specifically, the ability of a classifier to generalize to data points not used in training depends upon both the bias and variance of the classifier. Intuitively, *bias* refers to the ability of the classifier to fit the training data, i.e., a low-bias classifier gives accurate results on the training data. *Variance* refers to the flexibility of a classifier, i.e., a high-variance classifier changes significantly even with a slight change in the training data. The *bias–variance trade-off* in machine learning states that a classifier with low bias usually has high variance and vice versa. Ensemble models reduce generalization error by reducing the bias and/or variance in the classifier.

In the basic Bagging approach, a (sub)sample of the given training data is drawn (with replacement), and a classifier f_i is learned using this sample. Repeating the process k times produces k classifiers $f_1 \ldots f_k$. Given a new data point $\mathbf{x}$ to classify, these classifiers produce k labels $f_1(\mathbf{x}) \ldots f_k(\mathbf{x})$, and the final class for x is decided based on the majority.

Bagging does not have an effect on the bias of a classifier. To reduce bias, Boosting methods are used on a classifier. Boosting methods combine multiple *weak* classifiers into a combined classifier. Weak classifiers are those that may

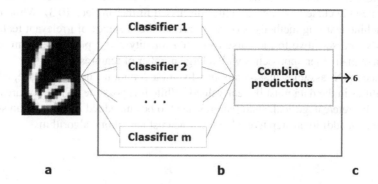

a   b   c

Fig. 2.4 Ensemble classifiers combine the labels from several other classifiers (e.g., by majority) on **a** a common given input (*left*) to produce a label **c** (*far right*) for the same input

not fit the training data completely. That is, they have high bias and are simple classifiers. The idea in boosting is to add several such simple classifiers to make a combined classifier that has low bias. The most popular version of Boosting is the AdaBoosting algorithm, a binary classifier with outputs of $+/-$. AdaBoosting is an additive model that sequentially adds classifiers so that the classifier added in iteration k is likely to correct the errors made by the classifier on the training data in iteration $k-1$. To do this, AdaBoosting maintains a weighted training dataset, where weights roughly correspond to the importance of choosing the instance to train the next classifier in the ensemble. Specifically, in each iteration of Adaboosting, the samples are re-weighted based on the errors made on the training data in the previous iteration. A classifier is learned in each iteration from a sample of the training data where the sampling is based on the weights, i.e., larger weighted instances are more likely to be chosen in the training sample. If $f_1 \ldots f_k$ are k classifiers learned in k iterations, they can be combined into a single classifier as follows:

$$f(x) = \text{sign} \left(\sum_{m=1}^{M} \alpha_m f_m(x) \right), \tag{2.3}$$

where each $f_m(x)$ is assumed to output a 1 or -1 value (assuming a binary classification) and α_m is a real-valued weight that encodes the importance of the mth classifier in the ensemble. The importance of the classifier added to the ensemble in this iteration is given by

$$\alpha_m = \frac{1 - err_m}{err_m},$$

where err_m is the error in the mth iteration. Thus, a smaller value of err_m implies that the classifier in the mth iteration has a larger weight in the overall ensemble. The training instances are re-weighed using an exponential re-weighting. Specifically, for all the correctly classified instances, their previous weight is multiplied by $\exp^{-\alpha_m}$ (which reduces the weight), and for the wrongly classified instances, their previous weight is multiplied by $\exp^{\alpha_m}$ (which increases their weight). It can be shown that this approach indeed reduces bias.

Example 2.8 (Ensemble Models for MNIST) One can combine several decision trees (defined in Sect. 2.1) to get an ensemble model for MNIST classification. Table 2.4 shows the precision, recall, and F1-score (described in Sect. 2.4) on a test dataset (chosen to be 25% of the data) for increasing the number of decision trees that are bagged together. Table 2.5 shows the same scores as the number of decision trees in the AdaBoosting ensemble is increased. Boosting the number of classifiers significantly improves performance since it reduces bias in the model, as compared to bagging that only reduces variance. □

Table 2.4 Performance of various ensemble solutions for the [**CharRC**] problem on the MNIST dataset increasing the number (N) of decision trees in the Bagging ensemble

N	Precision	Recall	F1-score
10	0.94	0.94	0.94
25	0.95	0.95	0.95
50	0.95	0.95	0.95
100	0.96	0.96	0.96

Table 2.5 Performance of various ensemble solutions for the [**CharRC**] problem on the MNIST dataset increasing the number of decision trees in the AdaBoosting ensemble (N)

N	Precision	Recall	F1-score
10	0.65	0.60	0.60
25	0.81	0.78	0.79
75	0.87	0.81	0.83
100	0.92	0.98	0.90

Practical considerations in the use of ensemble methods include the following:

- Bagging typically works better with unstable classifiers [4] such as decision trees and neural networks. In such classifiers, small changes to the training data affect the learned model significantly. A specialized form of Bagging is implemented in a random forest [4] (an ensemble of decision trees). Random forests are extremely powerful classifiers. A study based on more than 150 standard classification benchmarks showed that random forests outperform most other classifiers in a majority of benchmarks [9].
- Though Adaboosting was originally designed for weak classifiers, in practice, it can be used with all classifiers. Decision trees work especially well with boosting methods. A newer boosting approach called gradient boosting [10] is not only more flexible than the original Adaboosting approach, but it is highly scalable. An open source implementation of Boosted trees is available with the XGBoost library.

2.2.3 Support Vector Machines (SVMs)

Example 2.9 In an attempt to separate the data points from two classes shown in Fig. 2.5 by a line, the best line separator is a line that maximizes its distance from the data points in either class. The idea in support vector machine (SVM) classifiers is to try to learn such a separator. In some cases, a linear separator may be too restrictive, and therefore, using the idea of kernels (Sect. 4.1 gives more details on kernels), the SVM can learn more complex shapes of separators, as illustrated in Figs. 2.5 and 2.6.

Support vector machines (SVMs) are binary classifiers that learn a function such that the distance between the decision boundary of this function and data instances

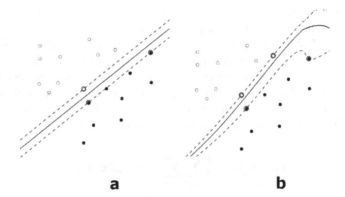

Fig. 2.5 (**a**) Linear SVM and (**b**) polynomial kernel decision boundary that separates data points of two classes (filled and empty circles). The encircled points are the support vectors, i.e., data points closest to the boundary

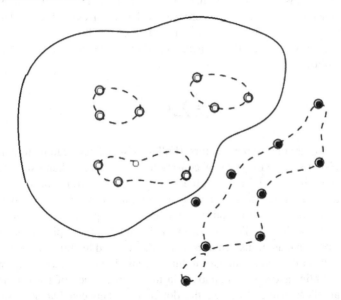

Fig. 2.6 RBF kernel SVM decision boundary that separates data points of two classes (filled and empty circles). The encircled points illustrate the support vectors, i.e., data points closest to the boundary

from either class is maximized. For example, if the training data contains 2 classes and the data corresponding to these classes can be separated by a linear decision boundary, the line may be very close to data points from either class, and the classifier is more likely to make mistakes on new data points that were not in the training data. On the other hand, if the distance between the line and the training data points on either side of the line is large, the classifier is more likely to assign the correct label on new data points. SVMs learn this type of decision boundary through a technique called max-margin optimization.

It turns out that learning a decision boundary in an SVM is fairly complex since it requires quadratic optimization. The idea behind this optimization procedure is to add *Lagrange coefficients* α_i corresponding to each data point. The optimization method solves a dual problem that computes the optimal values for these coefficients, from which the parameters for the classifier decision boundary can be derived. The objective function of the dual problem is

$$\max_{\alpha_i} \sum_{i=1}^{n} \alpha_i - \frac{1}{2} \sum_{i=1}^{n} \sum_{j=1}^{n} \alpha_i \alpha_j y_i y_j (\mathbf{x}_i \cdot \mathbf{x}_j), \tag{2.4}$$

where $\boldsymbol{\alpha} = \alpha_1 \ldots \alpha_n$ are the Lagrange coefficients, $\mathbf{x}_i \cdot \mathbf{x}_j$ is the dot product between the instances $\mathbf{x}_i$ and $\mathbf{x}_j$ in the training data, and $y_i y_j$ is the product of their class labels (assumed to be either 1 or -1). Importantly, in most cases, only a few Lagrange coefficients have nonzero values. These correspond to the data points that are those closest to the decision boundary, the so-called *support vectors*. The parameter vector $\mathbf{w}$ defining the decision boundary can be derived from the support vectors as follows:

$$\mathbf{w} = \sum_{i=1}^{l} \alpha_i y_i \mathbf{x}_i,$$

where $\alpha_1 \ldots \alpha_l$ are the support vectors. (Full technical details can be found in [7].)

The decision surface of SVMs is constrained to be linear when using the above max-margin optimization. That is, if the training data has two features, the decision surface is a line, for three features a plane and for n features, a $(n-1)D$ hyperplane. To learn nonlinear decision boundaries, the data can be pre-processed using *kernel* functions to enable SVMs. The idea in a kernel function is to implicitly add features (or dimensions) that are transformations of some selected features in the given data. Learning a linear decision boundary in this increased feature space is equivalent to learning a nonlinear decision boundary in the feature space of the original data. To learn an SVM using a kernel, the dot product between features, $\mathbf{x}_i \cdot \mathbf{x}_j$ in formula (2.4), is replaced by a general kernel function $K(\mathbf{x}_i, \mathbf{x}_j)$. There are two commonly used kernel functions, the polynomial kernel and the radial basis (RBF) kernel. The polynomial kernel combines features using a polynomial function, while the RBF kernel uses a Gaussian function.

Example 2.10 Table 2.6 shows the precision, recall, and F1-score (defined in Sect. 2.4) on a test dataset (chosen to be 25% of the data) for various types of kernels. The performance of SVMs with the RBF kernel yields the best performance in this case. □

Table 2.6 Performance of SVMs on the handwritten digit classification problem [**CharRC**] using different kernels to combine the pixels in the image to form higher-order features in pre-processing

Kernel	Precision	Recall	F1-score
Linear	0.967	0.967	0.967
Polynomial	0.967	0.953	0.956
RBF	0.982	0.982	0.982

Some practical considerations in using SVMs include:

- Practical SVM implementations generally perform a soft-margin optimization that contains a tunable hyper-parameter (typically called *cost*). Tuning this parameter allows misclassifications by the SVM in the training data when the training data cannot be separated by the decision boundary.
- SVMs are not easily explainable especially when using nonlinear decision boundaries. The effect of a single feature cannot easily be mapped to the output since features are combined with each other.

2.2.4 Neural Networks (NNs)

How do humans learn to recognize handwritten decimal digits, such as those in MNIST images? Most likely, the neurons in our brains are specialized in recognizing specific low-level features (such as line segments in certain orientations) and then combining them together into more complex patterns (such as curves) resulting in a full image, as discovered by Nobel prize winners, neurophysiologists D. Hubel and T. Wiesel in visual processing in cats' brains (*braintour.harvard.edu/archives/portfolio-items/hubel-and-wiesel*). A neural network (NN) is an artificial model that tries to emulate this process.

Example 2.11 Figure 2.7 illustrates a neural network. It typically consists of many processing units (akin to neurons) that are connected to each other by synaptic-like connections, each of which plays a tiny part in processing the image. The network is usually organized in layers, where previous layers pass on their outputs to neurons in a higher layer, which combines them to discriminate increasingly complex patterns in the input features. The final layer (usually a single neuron) gives an output that labels the actual classification of the original input vector. □

Neural networks were originally inspired by the working of the mammalian brain. Specifically, just like the human brain performs complex tasks using signals from an inter-connected network of neurons, neural networks learn functions through a distributed network of computational units. Each individual unit, referred to as a (artificial) *neuron*, is characterized by a certain number of *activation states* and performs a mathematical operation based on inputs given to it to produce an output that is forwarded to other units in the network, as specified by a

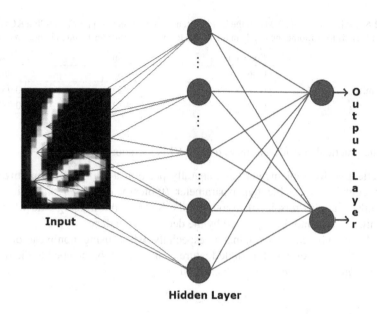

Hidden Layer

Fig. 2.7 A feedforward neural network (FNN) architecture consists of layers of neurons, including an input layer and an output layer. Inputs are fed to an input layer (e.g., pixels in an image, *left*); the network cascades these signals through any neuron in successive hidden layers (*middle*) refining feature extraction as it goes; finally, neurons in an output layer produce(s) a response in the output layer (e.g., digit 6 coded in binary (*right*))

directed graph called the *architecture* of the neural network. The most widely used architecture in neural networks is the *feedforward* architecture, where the units are arranged in layers and units in one layer are only connected to units in the layer next to it in the network. Each unit i in the neural network performs a seemingly insignificant but nonlinear operation that transforms inputs given to unit i into a single output based on weighted evidence net_i of all incoming units using a characteristic *activation function*, i.e., it determines how the individual unit in the neural network changes its activation. Specifically, each unit i has weights w_{ij} from other neurons j into it corresponding to each of its inputs. The inputs are multiplied by these weights, and the activation function is applied to this net input (the summation of the weighted outputs from other incoming neurons $x_j(t)$ at time t), i.e.,

$$A_i(t) = \sigma_i(net_i) = \sigma_j \sum_i w_{ij} x_j(t) \,,$$

where w_i is the weight for the ith input unit. The activation function σ_i on $net_i = \sum_i x_i w_i$ computes a single output for time $t + 1$. There are many possible types of networks depending on the type of activations and activation functions used to

produce the output. Most learning neural networks use *sigmoidal* units with the
activation function given by

$$\sigma(u) = \frac{1}{1 + e^{-u}},$$

where $u = net_i = \sum_j w_{ij}x_j$. Several other types of activations may be used,
including functions such as *tanh*, *RelU*, and *Softmax*. A *neural net* (NN) is thus
a complex structure consisting of an architecture, a matrix of synaptic weights, and
activation sets and functions for each neuron (usually a sigmoidal common to all).
Changing any of them, particularly, the weights in training, will change the network
because its responses to the same inputs will change accordingly.

Given a data point input, a FNN classifies the input by propagating the outputs
of the units from layer to layer. This is called *forward propagation*, where the input
features are clamped on the units of the *input layer*. The outputs from the input
layer are then forwarded to the units in the next layer (at the ends of the edges
pointing right in the architecture graph in Fig. 2.7). The neurons in layers that are
not connected to the inputs or output neurons are called *hidden neurons*. They may
in turn be organized into one or more hidden layers in the neural network. The
neurons feeding signals to no other neurons form the final *output layer*.

What confers upon FNNs their powerful and versatile learning ability is the
backpropagation algorithm. It is the most well-known learning algorithm that
usually works on most varieties of data science problems. For example for a
classification problem, the learning here is to produce the weights for all units
in the neural network such that the output answers given by the neural network
match the true classes in the data points in the training subset of data points as
closely as possible. Specifically, if $y_1 \ldots y_n$ are the labels for the n instances in the
training data and $o_1, \ldots, o_n$ represents the outputs of the neural network (given the
weights for the units in the neural network), backpropagation performs the following
optimization:

$$\mathbf{W_0} = \arg\min_{w_{ij}} \frac{1}{2} \sum_{k=1}^{n} (y_k - o_k)^2, \tag{2.5}$$

where $\mathbf{W} = [w_{ij}]$ represents the weight matrix among units j, i in the neural
network. Note that Eq. (2.5) assumes a single unit in the output layer for simplicity,
i.e., the neural network outputs a single value for each input instance. (In general,
the neural network can have multiple output neurons, but the same procedure is
applied to all of them.)

Backpropagation (BackProp) is a typical gradient descent algorithm (as
described in Sect. 11.3) if the activation function is differentiable (e.g., of sigmoidal
type). It solves the optimization problem for the objective function in Eq. (2.5). In
gradient descent, starting with a random initialization of weights, for each input
instance $\mathbf{x}_k$ in the training data, weight updates are roughly proportional to their

effect on the output o_k the network generates for that instance. Bigger weights tend to have a bigger effect on the outputs, and so more guilt is assigned to them for a wrong answer, i.e., they change more. Intuitively, for an input instance, if the neural network correctly classifies this instance, then the weights remain unchanged. But for an incorrect classification, BackProp modifies the weights so that if the same instance is encountered by the neural network again, the output of the modified network will be closer to the true label for that instance. The weight update for the output layer is slightly different from the weight update for the hidden layers. Assuming that all the units are sigmoidal units, the equation for updating the weight (w_{kj}) from unit j in the output layer is

$$\Delta w_{kj} = \eta(y_k - o_k)o_k(1 - o_k)I_j,$$

where η is a small constant (the *learning rate*), o_k is the output of the neural network for the k-th instance in the training data (thus, the difference $(y_k - o_k)$ is the error from the true class label incurred by the output of the neural network for the kth training data instance), and I_j is the value coming into the jth input of the output unit. Note that if $(y_k - o_k) = 0$, then the weights remain unchanged, i.e., the neural network accurately estimates the label for the kth instance with its current weights. Recursively, for a *hidden unit j* now, its weight update depends upon the units that it is connected to in the layer immediately above it. Intuitively, if $downCone(j)$ refers to all the units whose net input includes the output of j, the unit can influence the output of the overall neural network only through downstream units in $DownCone(j)$. The weight updates can reflect this influence as

$$\delta_j = o_j(1 - o_j) \sum_{k \in DownCone(j)} \delta_k w_{kj},$$

where o_j is the output of j and w_{kj} is the weight of the synaptic connection between j and one of its downstream units k. A larger weight implies a greater influence that j has through this downstream units. The update for hidden unit jth's input weights is then

$$\Delta w_{ji} = \eta \delta_j I_{ji},$$

where η is a small constant called the *learning rate* and I_{ji} is the value coming into the ith input for the jth hidden unit. In practice, the learning rate η can be finetuned to an optimal value based on experimental results. A very large value might cause the weights to fluctuate rapidly and never converge, while a very small value might result in slow convergence of the weights. Further, in practice, the weights may only be updated with a net change once for every *batch* of inputs in an epoch to reduce the running time of learning. A run of backpropagation where all the input data has been processed is termed an *epoch*. In real-world data, a neural network may require hundreds of epochs before the weights converge to an acceptable solution. (A full derivation and more details of the BackProp algorithm can be found in [20].)

Table 2.7 The performance of a neural network for [**CharRC**] improves with the size of the hidden layer since they are learning higher-level features from the pixels. Performance peaks out at 20 nodes for this choice of training data

N	Precision	Recall	F1-score
5	0.880	0.880	0.880
10	0.940	0.940	0.940
15	0.964	0.964	0.964
20	0.971	0.971	0.971
25	0.971	0.971	0.971

Example 2.12 (Neural Networks for MNIST) One can apply neural networks to the [**CharRC**] problem by considering each pixel as a feature in the input layer. Each pixel is then connected to all nodes in one hidden layer. The hidden layer neurons learn more complex, higher-level features of digit representations by combining pixels from the images into abstract features. Finally, the hidden layer is connected to the output layer to produce a class for the input image. The performance of neural networks for different hidden layer architectures is shown in Table 2.7 using precision, recall, and F1-score (defined below in Sect. 2.4) on a test dataset (chosen to be 25% of the data) as the number of nodes in the hidden layer progressively increases. The performance improves as the number of nodes in the hidden layer increases since it is learning higher-level features from the pixels. Once all the useful information has been represented, the performance peaks out at a certain point (20 nodes in this example), and adding neurons to the hidden layer does not contribute any significant performance improvements.   □

Practical considerations for neural network learning include:

- Neural networks are perhaps the most powerful and versatile among machine learning algorithms. A famous theorem discovered independently by several groups shows that neural networks are universal function approximators for continuous functions [11, 18], even after unbounded iteration as dynamical systems [12], i.e., a neural network, even with a single hidden layer, always exists to compute any continuous function on a given bounded region to any given arbitrary degree of accuracy. However, this does not always mean that neural networks produce the best approximations because large neural networks could require an infeasibly large training dataset or training time for backpropagation to learn appropriate weights.
- Overfitting is a common problem associated with neural networks. Overfitting occurs when the neural network fits the training data accurately (memorizes the data) but fails to generalize to data not used in training. One way to avoid overfitting is to regularize the neural network by forcing it to learn smaller valued weights by placing a penalty on large weights in Eq. (2.5). Recently, other approaches have been developed to avoid overfitting, such as *dropout*. The main idea in dropout is to force the neural network to learn simpler functions by removing units from the network at random.

- When the number of hidden layers is large (more than 2), the neural network is typically called a *deep network* [14]. Deep networks have obtained the state-of-the-art results in several applications such as image understanding, natural language understanding, and game playing. One of the issues with applying backpropagation to deep networks is that the weight updates tend to become 0 as the number of hidden layers increase, a problem referred to as the *vanishing gradient* problem. Therefore, variants of backpropagation algorithms have been developed to learn deep networks [17]. Further, since backpropagation and its variants use several matrix operations, they can be efficiently implemented in specialized hardware called GPUs to obtain significant improvements in the speed of large deep learning networks.
- Another problem with neural networks is that it is hard to explain/interpret results from the neural network, sometimes referred to as the *credit assignment* problem. Specifically, the hidden layers in the neural network transform the original features in the data into a representation that acts as a blackbox and makes it hard to understand or rationalize why it is producing the results it does. Further, since the final output from a neural network is based on a series of such transformations layer after layer, particularly in a deep network with several hidden layers, the network appears to be a black box harder to *interpret* by a human observer trying to make sense of the responses it is putting out. Interpreting the results in neural networks continues to be a highly active area of research that is vital to their use in real-world applications (Sect. 10.4 further discusses this matter).

2.3   Machine Learning Solutions: Unsupervised

Labels in the features in a dataset are extremely useful to find solutions to a problem using supervised learning methods, as described in the previous section. Naturally, answers to specific instances of a problem are not always available in the real world. For example, in a clustering problem, a label for a data point requires solving the problem holistically for *all* data points before it can be given, so it is impossible to include it in a dataset that one may hope to use to obtain a solution. Another family of methods in the area of unsupervised machine learning become useful to address such problems. Thus, the inputs to the algorithm are simply the data points without the labels. The only recourse left for unsupervised learning is to aim to discover regularities in the form of *hidden patterns* in the data in order to solve a problem. The goal of this section is to describe some of these methods, including the most popular ones, k-Means and Gaussian mixture models.

Example 2.13 By removing the labels in the Iris dataset (described in Sects. 1.1 and 11.5), a clustering problem arises. Ideally, the clustering would place instances of the same variety of Iris flower within the same cluster. To make the illustrations easy to visualize, just two features (the sepal width and sepal length) are considered in

the discussion in this section, although the methods can work with the full dataset
in higher dimensions just as well.   □

There are two main types of clustering:

- *Hard clustering* is ideal clustering, i.e., it requires a partition of the data points
  so that each instance belongs to exactly one part (or cluster).
- *Soft clustering* allows a fuzzy partition of the data points, where an instance is
  assigned to belong in more than one cluster (i.e., the parts in the partition overlap)
  or even perhaps no cluster at all (i.e., the parts do not exhaust the whole feature
  space), or both.

2.3.1 Hard Clustering

Example 2.14 The most intuitive way of assigning data points to clusters follows
up on the idea of a Voronoi diagram (defined in Sect. 1.1), i.e., to select a number
of *centroids* and assign the points to the nearest centroids. The problem is that there
are no centroids to begin with. Even if some initial centroids could be selected at
random, how can one ascertain that the clustering is of good enough quality? One
could choose as centroids the midpoints (i.e., the component-wise average of the
points) in a cluster and so require a Euclidean space where these operations can be
performed. Figure 2.8 shows such a clustering for the Iris dataset. In this case, the
assumption is that data points that are close to each other in terms of the Euclidean
distance are similar to each other and should be placed in the same cluster and vice
versa. It thus appears necessary to have some criterion to assess the quality of a
clustering to decide the assignment is good enough or refine it.   □

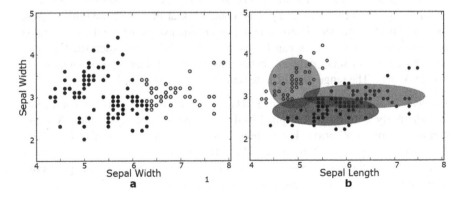

Fig. 2.8 Clustering of the Iris data. The three optimal clusters are indicated by gray shades. (**a**)
k-Means clustering with three clusters based on sepal features; (**b**) Soft clustering by Gaussian
mixture clustering

The most popular and elegant algorithm (due to its simplicity) for hard clustering is the *k-Means* algorithm for cases where the measure of similarity defining the problem is a Euclidean distance and a desired number of clusters k are given. Specifically, if the input is the usual data matrix $[\mathbf{x}_1 \ldots \mathbf{x}_n]$, where each data instance is a point in a Euclidean space (i.e., all its features are real numbers), k-Means starts with a random initialization of k cluster centroids and iteratively performs the following two steps aimed to improve performance until convergence:

1. Given the current centroids, assign each data point to the cluster determined by its closest cluster centroid.
2. Given assignments of all points to their clusters, update the cluster centroids to the mean of all its data points.

Convergence is achieved when the centroids no longer change. Figure 2.8 shows the results of k-Means applied to the Iris dataset.

To assess the quality of the solution thus provided, consider the functional Δ given by

$$\Delta = \min_{\{\mathbf{c}_j\}} \left\{ \sum_{i=1}^{n} |\mathbf{x}_i, \mathbf{c}_{j_i}|_2 \right\}$$

for the ℓ_2 distance, where $\mathbf{c}_{i_j}$ represents the cluster to which $\mathbf{x}_i$ is assigned, $\mathbf{c}_j$ is the cluster center of a cluster j, and $|*, *|_2$ is the ℓ_2 distance. Δ is a measure of the *within cluster variation* for a choice of k centroids $\{\mathbf{c}_j\}$. It has been shown [3] that k-Means converges to a local minimum of this aggregated measure of how close the data points are to their centroids across all clusters, so k-Means does solve an optimization problem. However, convergence is not guaranteed, and when it does converge, the clusters may not be guaranteed to be optimal across all possible choices of centroids.

Inter-cluster variance minimization is re-assuring, but the value itself tells little about the quality of the clustering. One could look at the results and assess them visually, but that becomes impossible in higher dimensions, where the challenging problems lie. Other metrics can be used, such as *silhouette* (defined below in Sect. 2.4). A larger score indicates better clustering of similar images compared to smaller scores. The range of silhouette scores is the interval $[0, 1]$.

Example 2.15 (k-Means on the MNIST Dataset) Each pixel is considered a feature, and similarity is based on the pixel RGB (or grayscale) values. Table 2.8 shows silhouette scores for various k, the number of clusters. As seen in this table, as we approach the true number of clusters (here 10), the silhouette score progressively increases indicating that k-Means is detecting true patterns in the images corresponding to each digit and is therefore able to cluster them optimally.   □

Table 2.8 Performance of k-Means (*second column*) and Gaussian Mixtures Models (GMMs) (*third column*) on the MNIST dataset for various numbers of clusters (k). As k nears the optimal value 10, the silhouette score progressively increases to an optimal value of quality for k-Means (still far from the ideal) and then increases further for GMMs

k	Silhouette k-means	Silhouette GMM
1	0.116	0.100
3	0.118	0.080
5	0.137	0.040
8	0.157	0.060
10	0.159	0.167

Some practical considerations when using k-Means clustering include:

- The clustering generated using k-Means depends upon the initial cluster centroids. More sophisticated initialization methods than random initialization are used in algorithms such as k-Means ++ [1].
- k-Means works well when the optimal clusters are *globular*, i.e., are roughly spherical-shaped. Likewise, k-Means works well when the clusters have similar sizes and densities, i.e., have roughly the same number of points in each cluster with a similar spread of points.
- With nonglobular shapes, k-Means tends to separate clusters poorly. For nonspherical clusters, a variant called kernel k-Means can be used where kernel functions are applied to increase dimensionality of the data before clustering. Alternatively, if the clusters are diamond-shaped, a different metric such as ℓ_1 (defined in Chap. 1 and illustrated in Fig. 1.8) could be used.

2.3.2 Soft Clustering

Example 2.16 In clustering the Iris dataset, if a certain flower data point's features put it into two different clusters, instead of forcing that instance into a single cluster, one can determine how likely is an instance to belong to a cluster. The right panel of Fig. 2.8 shows the clusters formed using a Gaussian Mixture Model (GMM) for the Iris data. The three shaded regions illustrate the shape of the Gaussians, and so these distributions will overlap, as shown. For every flower in the dataset, one can now determine how likely it is to be part of the 3 clusters, and therefore, it is perfectly feasible that a flower that lies at the intersection of two clusters is equally likely to be a member of either cluster. □

When the optimal clusters are uncertain, i.e., each data point can belong to multiple clusters, instead of randomly choosing a cluster for a data point, probabilistic methods can be used. The most well-known approach for probabilistic clustering is the *Gaussian Mixture Model* (GMM)-based clustering. GMMs are probability distributions using multiple Gaussians with some designated means. In the case of clustering, each cluster is represented by a Gaussian in the mixture. The expectation–maximization (EM) algorithm (described in Sect. 11.3.2) is used to learn the parameters of the GMM. Specifically, if there are k clusters in the data, a GMM distribution is defined as

$$P(x|\Theta) = \sum_{i=1}^{k} \alpha_i P_i(x|\theta_i),$$

where $P_i(x|\theta_i)$ is the ith Gaussian probability distribution and α_i is its weight. The ith Gaussian is defined by mean and co-variance matrix parameters $\theta_i = (\mu_i, \Sigma_i)$, i.e., its probability density function is

$$P(\mathbf{x}|\theta_k) = \frac{1}{(2\pi)^{p/2}|\Sigma_k|^{1/2}} \exp^{-\frac{1}{2}(\mathbf{x}-\mu_k)^{\mathsf{T}} \Sigma^{-1}(\mathbf{x}-\mu_k)},$$

where p is the number of features in $\mathbf{x}$. The parameters θ_i can be learned using the EM algorithm. EM starts by assigning random values for the Gaussian parameters in the mixture and iteratively updates them until they converge by repeatedly performing the following two steps:

- *Expectation step*
  The probability of each data point is computed for each of the Gaussians in the mixture using the current Gaussian clusters. This step performs the soft clustering.
- *Maximization step*
  The Gaussians are re-parameterized based on the probabilities of the data points computed in the expectation step using a method called *max-likelihood estimation* (also described in Sect. 11.3.2).

Expectation–maximization is repeated until all the parameters for all the Gaussian distributions have converged.

Example 2.17 (GMMs for MNIST) Table 2.8 also shows the silhouette scores as the number of Gaussians (N) in the GMM varies. As seen in this table, the results are less intuitive than k-Means since increasing the Gaussians initially reduces the silhouette score that seems counterintuitive. One reason for this could be that since GMMs perform soft clustering, for similar looking digits (e.g., 1 and 7), the probability of that digit belonging to several classes is somewhat similar, and thus it is hard to make a distinction as to which class it truly belongs to. However, the silhouette score increases indicating that as the sufficient number of Gaussians in the GMM (in this case 10 digits) is reached, each Gaussian is modeling images corresponding to a single digit.   □

Some practical considerations when using Gaussian mixtures models include:

- GMMs are very expressive and can represent a wide range of distributions. In fact, they are considered as universal approximators, i.e., they can represent any type of distribution [27].
- The EM algorithm is not guaranteed to find the optimal clusters.
- While the basic GMM clustering algorithm pre-specifies the number of clusters (i.e., it fixes the number of Gaussian distributions in the mixture), there are more advanced variants of nonparametric Gaussian Mixture Models that infer the optimal number of clusters in the data [6].
- A well-known application of GMM clustering is in topic modeling where the task is to infer the number and topics in a text document [2].

2.4 Controls, Evaluation, and Assessment

Two tasks are critical in solving a well-defined problem in data science, namely, the methods to find solutions and the evaluation and assessment of their quality [21]. This section describes methods and metrics to quantify and evaluate the quality of solutions. Methods include training, testing, and cross-validation. Metrics include accuracy, precision (specificity), recall (sensitivity), and clustering metrics. Such methods are required not only in the training phase of the solution or model but more importantly in the testing and cross-validation phases of a solution development cycle. In addition, an assurance that the solution will work well in the production environment is also desirable.

Example 2.18 The Netflix movie recommendation problem can be regarded as a classification problem. A good movie recommendation system should provide a user with a list containing movies that s/he is most likely to like. Therefore, a solution not only needs to identify whether the user is going to like (with varying degree of preference) a movie, but also predict the corresponding ratings so that the system can figure out the top 5 or so to recommend the user. Accuracy does not appear to be an appropriate metric to quantify performance since there are no objective labels associated with a rating, and they are viewer-dependent. What would be an appropriate metric to use? Should it use all the data points or only some of them to get a sense how the solution will perform on unknown future data points?   □

2.4.1 Evaluation Methods

Generally, a dataset is split into three subsets, i.e., a training set, a testing set, and a validation set (already mentioned in Sect. 1.3). A *training* dataset is a set of data points used to train a model by fitting a set of relevant parameters. At the end of the training phase, a fitted model is obtained that can then be assessed using a *testing* dataset, i.e., a set of data points left out during the training phase, to determine the model's quality. Finally, a *validation dataset* is a set of instances used to tune the hyper-parameters for the production environment [23]. The distinction between the testing and the validation phase is less sharp since both require data points left out during the training phase. However, a distinction can be made [23] by defining a validation set as a set of points used to tune the parameters of a classifier (for example, the number of hidden units in a neural network), whereas the testing set is used to assess the performance of a fully specified classifier. Several quantitative metrics used to measure the performance of a ML model in each phase are summarized in Table 2.10. For low-scale applications, just training and testing sets are used to zoom into a solution.

A more robust (hence very popular) approach is k-fold *cross-validation* ($k \geq 2$). In this approach, the dataset is split into, say $k = 5$ parts, and a model is then trained using four partitions and tested on the remaining partition, recording the performance score on the latter partition. The process is repeated s number of times (16 is a common number to establish a significance for a sample of small size), and the average of these cross-validation scores is used as a metric to assess the quality of the model.

2.4.2 Metrics for Assessment

In terms of specific scores for a given dataset, there are a number of choices, each appropriate for various kinds of problems and cases. They are summarized in Table 2.10. In the simplest case of a classification problem, the performance of a solution (called a *classifier*) can be measured in various ways. The simplest one is well known.

> The *accuracy* of a classifier M on a dataset **X** is the ratio $hits/misses$, where a *hit* (*miss*)
> is data point **x** for which M's answer does (does not, respectively) agree with the label in **x**.

A binary classifier M for a classification problem with two categories T and F (so that $\Omega = T \cup F$, say T being the class of true interest) may *miss* in two different ways: by placing the data point from F into T (a so-called *false positive (FP)* , or vice versa (a so-called *false negative (FN)*) and ditto for F. A *hit* will place elements of T in T, i.e., the *True Positives (TP)* and *True Negatives* obtained from M satisfy $|T| = |TP| + |FN|$. The accuracy would then be $a = (|TP| + |TN|)/(|T| + |F|)$. If the classifier is not binary, hits and misses are calculated for elements in a class of interest T by considering F to be the union of the remaining classes as a binary classification.

Accuracy can be very misleading in lopsided classification problems where one class is disproportionally large compared to the others. A lazy classifier can simply place all elements in the large class and be assured high accuracy. In many situations, especially in clinical settings, where T is healthy patients and F is patients with some disease, the category of more interest is really F despite the fact it may be a relatively small class. In such a case, there are three other choices, depending on which of the two categories is more important, T or F.

The *recall* (also called *sensitivity*) of a classifier M on a dataset $\mathbf{X}$ is the ratio $TP/(TP + FN)$, i.e., the proportion of correct classifications out of the total number of *positive* classifications made (only considering the class T, which includes the FNs for the classifier).

The *precision* of a classifier M on a dataset $\mathbf{X}$ is the ratio $TP/(TP + FP)$, i.e., the proportion of correct classifications out of the total number of *positive* classifications made (only considering the class F, which includes the FPs for the classifier).

The *specificity* of a classifier M on a dataset $\mathbf{X}$ is the ratio $TN/(TN + FP)$, i.e., the proportion of correct classifications out of the total number of *negative* classifications made (only considering the class F, which includes the FPs for the classifier).

The *F1-score* of a classifier M is the geometric mean

$$F1 = \frac{1}{\frac{1}{\text{recall}} + \frac{1}{\text{precision}}} = \text{precision} * \text{recall}/(\text{precision} + \text{recall}).$$

These quantities are usually displayed together in a so-called *confusion matrix* (no doubt, to avoid confusion!), as illustrated in Table 2.9.

For prediction problems where the target (or loss) function f takes on numerical values, the error is measured by some ℓ_d (usually $d = 1$ or 2) distance between the predicted value and the observed value (or label), either on average or relative to the true value (Table 2.10).

The *root mean-squared error (RMSE)* is the average ℓ_2 distance between the model value and the true value, averaged across all data points, i.e.,

$$\text{RMSE} = \frac{1}{n} \sqrt{\sum_{i=1}^{n} (\mathbf{y}_i - \mathbf{x}_i)^2 / n}.$$

The *relative error (RE)* is the average ℓ_1 distance (absolute value) between the model value and the true value scaled to the true value, i.e.,

$$\text{RE} = \frac{1}{n} \sum_{i=1}^{n} |\text{Observed}_i - \text{Predicted}_i| / |\text{Observed}_i|.$$

Table 2.9 Confusion matrix for a classifier solution to a classification problem in relation to a category T and the remaining categories F (The brackets are actual values for a given problem and dataset)

	Predicted	
Observed	True	False
True (in T)	[TP]	[FN]
False (not in T)	[FP]	[TN]

Table 2.10 Common quantitative metrics to assess the quality of a data science solution M [13, 15, 23, 24, 26] for a dataset $\mathbf{D}$ of size n and data points $\mathbf{x}_1, \ldots, \mathbf{x}_n$

Metric	Definition
Classification	
Accuracy	$a = (TP + TN)/(TP + TN + FP + FN)$
Recall (sensitivity)	$r = TP/(TP + FN)$
Precision	$p = TP/(TP + FP)$
Specificity	$p = TN/(TN + FP)$
F1-score	$F1 = pr/(p + r)$
Prediction	
Root mean-squared error (RMSE)	$\text{RMSE} = \sqrt{\sum_{i=1}^{n}(\text{Observed}_i - \text{Predicted}_i)^2/n}$
Relative error (RE)	$\text{RE} = \frac{1}{n}\sum_{i=1}^{n} \lvert \text{Observed}_i - \text{Predicted}_i \rvert / \lvert \text{Observed}_i \rvert$
Clustering $\mathbf{D} = C_1 \cup \ldots \cup C_k$	
SSD	Total sum of squared distances of $\mathbf{x}_i$s from their centroids $\bar{\mathbf{x}}_i$: $\text{SSD} = \sum_{i=1}^{n} \text{SSD}_i(M) = \sum_{i=1}^{n} d(\mathbf{x}_i, \bar{\mathbf{x}}_i)^2$
k-Means—Elbow Method	Choose k that causes a sharp turn in SSD
Silhouette value of a point	$s(i) = (b(i) - a(i))/\max\{a(i), b(i)\}$, where $a(i) = \frac{1}{(\lvert C_{k_i}\rvert - 1)} \sum_{\mathbf{x}_j \in C_{k_i}, \mathbf{x}_j \neq \mathbf{x}_i} d(\mathbf{x}_i, \mathbf{x}_j)$, $b(i) = \min_{j \neq k_i} \frac{1}{\lvert C_j \rvert} \sum_{\mathbf{u} \in C_j} d(\mathbf{x}_i, \mathbf{u})$, C_{k_i} is the cluster given by M for $\mathbf{x}_i$ and $d(\mathbf{x}_i, \mathbf{u})$ is the distance between data points $\mathbf{x}_i$ and $\mathbf{u}$
Silhouette of a clustering	s is the average of the $s(i)$ values of all points $\mathbf{x}_i$ in $\mathbf{D}$
Silhouette score for k	s_k for k clusters is the average of all such s's
Silhouette coefficient	Of $\mathbf{D}$ is the maximum silhouette score s_k over all $1 \leq k \leq n$
Regression	
R-squared	$R^2 = 1 - (\text{RSS}/\text{TSS})$, where RSS is the sum of squares of residuals and TSS is the total sum of squares
Akaike's Information Criterion (AIC)	$\text{AIC} = 2k - 2\log(L)$, where k is the number of estimated parameters in the model M and L is the maximum value of the likelihood function for M
Bayesian Information Criterion (BIC)	$\text{BIC} = k\log(n) - 2\log(L)$, where L is the maximized value of the likelihood function of the model M and k is the number of parameters estimated by M

For clustering problems, there are several metrics to assess the quality of a solution. They focus on measuring the density of the clusters (average intra-cluster distance), as well as the separation between clusters (average inter-cluster distance), on the average. A popular combined measure is the silhouette.

The *silhouette value* of a point $\mathbf{x}_i$ in a clustering of a dataset
$\mathbf{D} = C_1 \cup \cdots \cup C_k$ is given by

$$Silhouette\ s(i) = (b(i) - a(i))/\max\{a(i), b(i)\},$$

where $a(i)$ is the average *cohesion* value of a data point $\mathbf{x}_i$ in its cluster C_{k_i} to all other data
points in the same cluster, measured by $a(i) = 0$ if $C_{k_i} = \{\mathbf{x}_i\}$, else by

$$a(i) = \frac{1}{(|C_{k_i}| - 1)} \sum_{\mathbf{x}_j \in C_{k_i}, \mathbf{x}_j \neq \mathbf{x}_i} d(\mathbf{x}_i, \mathbf{x}_j),$$

$b(i)$ is the minimum *separation* of the data point $\mathbf{x}_i$ with all data points in the closest cluster
measured by

$$b(i) = \min_{j \neq k_i} \frac{1}{|C_j|} \sum_{\mathbf{u} \in C_j} d(\mathbf{x}_i, \mathbf{u}),$$

and $d(\mathbf{x}_i, \mathbf{u})$ is the distance between data points $\mathbf{x}_i$ and $\mathbf{u}$.
The *silhouette* for the clustering solution is the average of the silhouettes $s(i)$ of its data
points $\mathbf{x}_i$. The *silhouette score* s_k for k clusters is the average of the silhouettes s for all
clustering with k clusters. The *silhouette coefficient* for a dataset is the maximum of the
silhouettes scores s_k across all possible values $1 \leq k \leq n$ (Table 2.11).

Once a performance metric is selected and computed on a test dataset, an
important question arises: *is this score good enough for a quality solution?* In
Example 2.18, if a recommender system got an average RE of 50%, can it be
deployed in a market? To make such a decision, a more thorough analysis should
be made considering several solutions, including competitors' performance. One
approach is to compare it to published performance scores for similar problems as
a baseline and decide accordingly.

Another approach involves running an *experimental control*. This approach is
common for decision problems whose solutions have a large impact on a person

Table 2.11 Use of assessment metrics in typical problems in data science (PwP: Possible with
Pre-processing)

Metrics	Classification	Clustering	Prediction
Accuracy	Ok if supervised	N/A (no labels)	PwP
Recall (Sensitivity)	Ok if supervised	N/A (no labels)	PwP
Precision	Ok if supervised	N/A (no labels)	PwP
Specificity	Ok if supervised	N/A (no labels)	PwP
F1-score	Ok if supervised	N/A (no labels)	PwP
Silhouette	PwP	Yes	PwP
R-squared	PwP	N/A (no response)	Yes
Akaike's Information Criterion (AIC)	PwP	N/A (no response)	Yes
Bayesian Information Criterion (BIC)	PwP	N/A (no response)	Yes
Root mean-squared error (RMSE)	PwP	N/A (no response)	Yes
Relative error (RE)	PwP	N/A (no response)	Yes

or population. The simplest case is the well-known *placebo control*, where the effectiveness of a drug may depend on subjective self-reporting that may be misleading (the well-known placebo effect). A control requires changing a critical feature predictor (provide a placebo instead of the actual drug) and comparing the results with the complementary case where the predictor has the opposite value (e.g., the actual drug is administered), while *keeping everything else unchanged.* The difference in scores will be considered significant if the difference between the scores is larger than the standard error of the scores in the positive case (e.g., actual drug taken).

In summary, the decision whether a particular score is good enough really depends on the definition of the business problem being tackled (as defined in Sect. 1.4). Without that definition, it is impossible to make a decision. With it, there is still room for argument in terms of the impact of the solution or decision being implemented. A solution to be deployed into a market for production requires careful consideration of financial and/or other implications. A movie recommendation system that is going to be used by a handful of people would tolerate a medium score. If the number of viewers ranges in the millions, the impact may require much higher scores.

References

1. Arthur, D., & Vassilvitskii, S. (2007). k-Means++: The advantages of careful seeding. In *SODA '07: Proceedings of the Eighteenth Annual ACM-SIAM Symposium on Discrete Algorithms* (pp. 1027–1035).
2. Blei, D. M., Ng, A. Y., & Jordan, M. I. (2003). Latent Dirichlet allocation. *Journal of Machine Learning Research, 3,* 993–1022.
3. Bottou, L., & Bengio, Y. (1995). Convergence properties of the k-means algorithms. In G. Tesauro, D. Touretzky, & T. Leen (Eds.), *Advances in neural information processing systems* (Vol. 7). MIT Press.
4. Breiman, L. (2001). Random forests. *Machine Learning, 45*(1), 5–32.
5. Chen, C., Liu, Y., & Peng, L. (2019). How to develop machine learning models for healthcare. *Nature Materials, 18,* 410–414.
6. Christopher, M. (2006). *Pattern recognition and machine learning.* Springer.
7. Cristianini, N., & Shawe-Taylor, J. (2000). *An introduction to support vector machines and other Kernel-based learning methods.* Cambridge University Press.
8. Domingos, P., & Pazzani, M. (1997). On the optimality of the simple Bayesian classifier under zero-one loss. *Machine Learning, 29*(2), 103–130.
9. Fernández-Delgado, M., Cernadas, E., Barro, S., & Amorim, D. (2014). Do we need hundreds of classifiers to solve real world classification problems? *Journal of Machine Learning Research, 15*(1), 3133–3181.
10. Friedman, J. B. (2001). Greedy function approximation: A gradient boosting machine. *Annals of Statistics, 29,* 1189–1232.
11. Funahashi, K. I. (1989). On the approximate realization of continuous mappings by neural networks. *Neural Networks, 2*(3), 183–192.
12. Garzon, M., & Botelho, F. (1999). Dynamical approximation by recurrent neural networks. *Neurocomputing, 29*(1), 25–46.

13. Glantz, S. A., Slinker, B. K., & Neilands, T. B. (1990). *Primer of applied regression and analysis of variance*. McGraw-Hill Inc.
14. Goodfellow, I., Bengio, Y., & Courville, A. (2016). *Deep learning*. MIT Press.
15. Gideon, S., et al. (1978). Estimating the dimension of a model. *The Annals of Statistics, 6*(2), 461–464.
16. Hastie, T. J., & Tibshirani, R. J. (1986). Generalized additive models. *Statistical Science, 43*(3), 297–310.
17. Hinton, G. E., Osindero, S., & Teh, Y. W. (2006). A fast learning algorithm for deep belief nets. *Neural Computation, 18*, 1527–1554.
18. Hornik, K., Stinchcombe, M., & White, H. (1989). Multilayer feedforward networks are universal approximators. *Neural Networks, 2*(5), 359–366.
19. Kelleher, J. D., MacNamee, B., & D'Arcy, A. (2015). *Fundamentals of machine learning for predictive data analytics: Algorithms, worked examples, and case studies*. MIT Press.
20. Mitchell, T. M. (1997). *Machine learning*. McGraw-Hill.
21. Mount, J., & Zumel, N. (2019). *Practical data science with R*. Simon & Schuster.
22. Quinlan, J. R. (1986). Induction of decision trees. *Machine Learning, 1*, 81–106.
23. Ripley, B. D. (2007). *Pattern recognition and neural networks*. Cambridge University Press.
24. Rousseeuw, P. J. (1987). Silhouettes: A graphical aid to the interpretation and validation of cluster analysis. *Journal of Computational and Applied Mathematics, 20*, 53–65.
25. Schapire, R. E. (2013). Explaining AdaBoost. In *Empirical inference* (pp. 37–52). Springer.
26. Taddy, M. (2019). *Business data science: Combining machine learning and economics to optimize, automate, and accelerate business decisions*. McGraw Hill Professional.
27. Titterington, D. M., Smith, A. F. M., & Makov, U. E. (1985). *Statistical analysis of finite mixture distributions*. New York: Wiley.

Chapter 3
What Is Dimensionality Reduction (DR)?

Lih-Yuan Deng (iD)**, Max Garzon** (iD)**, and Nirman Kumar** (iD)

Abstract Solutions to problems require either assumptions on the target population or lots of data to train models that may help answer the questions. Our ability to generate, gather, and store volumes of data (order of tera- and exo-bytes, $10^{12} - 10^{18}$ daily) has far outpaced our ability to derive useful information from it in many fields, with available computational resources. Therefore, data reduction is a critical step in order to turn large datasets into useful information, the overarching purpose of data science. DR thus becomes absolutely essential in DS, particularly for big data.

3.1 Dimensionality Reduction

With the proper context for data science presented in the previous chapters, we can now proceed to the subject matter proper, dimensionality reduction (DR). This chapter presents a high-level overview of the various approaches to DR in the literature. To frame them in a conceptual framework, first a general and systematic definition of DR is required.

In the simplest case, a problem can be regarded as defined by a number of features, one of which is the so-called *response* (random) variable in statistics, or simply *target* feature in this book. The remaining features are the so-called *predictors* or *independent variables* in statistics, or simply *input* features in this book. A sample of these features from the corresponding population constitutes the data to be used to solve the problem. As discussed in Chap. 1, the sample $\mathbf{X}$ can be regarded as a table consisting of a number n of observations ($n > 1$) given by vectors $(\mathbf{x}_1, y_1), \ldots, (\mathbf{x}_n, y_n)$ where $\mathbf{x}_i = (x_{i1}, \ldots, x_{ip})$ is a covariate vector of input features for the ith observation and y_i is the target (response) variable. Without

L.-Y. Deng (✉)
Mathematical Sciences, The University of Memphis, Memphis, TN, USA
e-mail: lihdeng@memphis.edu

M. Garzon · N. Kumar
Computer Science, The University of Memphis, Memphis, TN, USA
e-mail: mgarzon@memphis.edu; nkumar8@memphis.edu

loss of generality one can assume that the target variable is of a scalar type and is not included in **X** because the same approach can be repeated for each scalar component in a multi-dimensional target vector. Thus the dataset can be represented as a matrix

$$
\mathbf{X}_{np} =
\begin{bmatrix}
x_{11} \cdots x_{1p} \\
x_{21} \cdots x_{2p} \\
\cdot \\
\cdot \\
\cdot \\
x_{n1} \cdots x_{np}
\end{bmatrix}
= [X_1, X_2, \cdots, X_p], \quad
\mathbf{Y} =
\begin{bmatrix}
y_1 \\
y_2 \\
\cdot \\
\cdot \\
\cdot \\
y_n
\end{bmatrix}, \tag{3.1}
$$

where the input matrix can be viewed as n row vectors $(\mathbf{x}_1, \mathbf{x}_2, \cdots, \mathbf{x}_n)$ of dimension p (the data points) or p column vectors $(X_1, X_2, \cdots, X_p)$ of dimension n (the features) and **Y** is the response column vector of dimension n. (To ease notation, column vectors like BY will be shown as transposes $(y_1, y_2, \ldots, y_n)'$ of row vectors in the sequel.)

Let $\bar{\mathbf{X}}$ be the matrix of $n \times p$ with each row as (row) sample average of the matrix product $\mathbf{X} = \mathbf{1}'\mathbf{X}/n$, where $\mathbf{1} = (1, 1, \cdots, 1)'$ is an n-dimensional column vector of 1s. The sample variance–covariance matrix is

$$
\mathbf{S} = \frac{1}{n-1}(\mathbf{X} - \bar{\mathbf{X}})'(\mathbf{X} - \bar{\mathbf{X}}), \quad \bar{\mathbf{X}} = \frac{1}{n}\mathbf{1}\mathbf{1}'\mathbf{X}. \tag{3.2}
$$

In general, the overarching goal of DR is *to find lower dimensional representations of data that preserve their key properties for a given problem.* Historically, the classical technique to DR was principal component analysis, commonly referred to as PCA. Other techniques were eventually developed to include feature/variable selection (particularly including targets) rather than feature extraction. In general, choosing between feature extraction with PCA or just feature/variable selection depends mostly on the problem being solved. DR is most effective for big data with a large number of input variables that are correlated with each other.

The classical statistical approach assumes that the solution to a given problem is some function f of p covariates in **X** that produces the nD *response vectors* **Y** (containing the responses for all the n input feature vectors in the data), where p is too large to efficiently build a good approximation of f. Typically, **X** may have a high correlation among some of the columns, or some columns may have a nonlinear relationship. Therefore, one can first attempt to reduce the number of features to a much smaller number k ($k \ll p$). This reduction can be viewed as a transformation Ψ of the original dataset **X** into another matrix $\mathbf{X}^*$ with k columns,

$$
\Psi(\mathbf{X}) \equiv \mathbf{X}^* \equiv [X_1^*, X_2^*, \cdots, X_k^*]. \tag{3.3}
$$

This reduced feature set could then be used to solve the problem of determining the response from the features as a relation between $\mathbf{Y}$ and $\mathbf{X}^*$ given by

$$\mathbf{Y} = f(\Psi(\mathbf{X}^*)) + \epsilon, \tag{3.4}$$

where ϵ is a random error term and $f(\cdot)$ is the response function.

Ψ could be a *selection* from the original (given raw) features in X or could be a combination of them into other new *derived* (or *abstract*) features. Generally, there are two more cases to consider, reductions $\Psi(\cdot)$s which would only involve the matrix $\mathbf{X}$, not the response $\mathbf{Y}$, and reductions that do. The first kind of reduction originates in statistics, where it is considered a kind of *statistical inference*. On the other hand, feature selection/*variable selection* refers to methods for identifying fewer "optimal" in a feature subset yielding a small error rate ϵ.

Example 3.1 A linear mapping is most useful when some of the column vectors in $\mathbf{X}$ are linearly correlated with each other. In this case,

$$\Psi(\mathbf{X}) = \mathbf{XC}, \tag{3.5}$$

where $\mathbf{C}$ is a matrix representing variable selection and/or DR. Various choices for the matrix $\mathbf{C}$ in the model (3.5) produce several cases:

1. First, when $\mathbf{C} = \mathbf{I}$, the p-dimensional identity matrix, it is reduced to the full set of features $\mathbf{X}$ and a model $\mathbf{Y} = f(\mathbf{X}) + \epsilon$.
2. When $\mathbf{C}$ is a subset of columns in $\mathbf{I}$, it is reduced to the usual sub-model with only a few columns/variables that can be selected by various methods (e.g., entropy-based, as discussed in Chap. 6).
3. The PCA method can be considered a special case with $\mathbf{C}$ being a weight matrix consisting of a certain column combination of features in $\mathbf{X}$ that maximizes the variation of the data in $\mathbf{X}$.

□

Most of the classical statistical model building and machine learning methods are quite inefficient with a huge number of input variables. Applying an exhaustive search of the space for feature/variable selection is computationally intractable for all but the smallest of data sets, especially when the number of features p is large. Because of the complexity of big data, it is often necessary to use DR techniques before attempting to conduct statistical inference or solve a problem.

In addition to PCA and its variants, several other powerful DR techniques will be discussed throughout the remainder of this book. A summary of the methods is shown in Table 3.1, followed by a high-level overview of them in the following sections. Other efficient and powerful search algorithms in the literature for variable selection will also be introduced in Chap. 8.

Each of DR methods briefly described below is designed to maintain certain aspects of the original data and, therefore, may be appropriate for one task but inappropriate for another. In addition, most methods have parameters to tune and/or

Table 3.1 Dimensionality reduction methods

Method	Description	Section
Principal Component Analysis (PCA)	Creates key principal components capturing maximum data variation as combinations of orthogonal linear input variables	4.1
Kernel PCA (KPCA)	Generalizes of PCA to other possible weights for the distance variation	4.2.1
Independent component analysis (ICA)	Decomposes the data into several components for the purpose of object separation or clustering	4.2.2
Multi-Dimensional Scaling (MDS)	Given the distance between pairs of objects in a set, MDS places each object into a lower dimensional representation such that the between-object distances are (mostly) preserved	5.2.1
Isometric Mapping (ISOMAP)	Builds a mapping preserving a distance measure defined over a lower dimensional manifold that can fit the between-data distances reasonably well	5.2.2
t-Stochastic Neighbor Embedding (t-SNE)	Improves the probabilistic approach SNE transforming objects given by high-dimensional vectors or by pairwise dissimilarities, into a lower dimensional space preserving neighbor identities	5.2.3
Conditional Entropy	Selects features/input variables by minimizing conditional entropy, or equivalently, by maximizing mutual information or information gain	6.2
Iterated Conditional Entropy	Selects features/input variables by minimizing conditional entropy on previous choices iteratively	6.3
Conditional Entropy of Targets	Selects features/input variables by minimizing conditional entropy of the target/response variable	6.4
Reduction by Genomic Signatures	Uses the pointwise hybridization pattern exhibited by a dataset encoded as DNA sequences to noncrosshybridizing (nxh) DNA chips designs based on the deep structure of DNA spaces	7.3
Reduction by Pmeric Signatures	Uses the barycentric coordinates of a set of DNA sequences in a DNA space encoding the points in a dataset in the convex hull of the Euclidean encoding of their centroids	7.4

make specific assumptions. Therefore, the quality of the model or solution to a problem may strongly depend on their tuning, with the added complexity to DR methods.

3.2 Major Approaches to Dimensionality Reduction

This section presents a summary review of major dimensionality reduction (DR) methods for datasets and problem solutions subject of this book. Supervised methods require labeled data containing the expected answers, in addition to a given set of features describing a data point, whereas unsupervised methods just require the feature vector specifying the data point. The descriptions focus on practical aspects of these solutions related to solving data science problems, with further references to other sources for a more comprehensive treatment of the technique(s) as appropriate.

3.2.1   Conventional Statistical Approaches

This section describes major statistical methods for DR. To study whether a feature can help in solving a data science problem, most statistical models formulate the problem into statistical metrics such as probabilities mass/density functions. (Sect. 11.1 contains a description of basic probability and statistical background.) By assessing these metrics, a practitioner can select important features, describe the relationship among variables, make inferences, and classify, predict, or cluster future events.

From a statistical perspective, a most important metric of the data is its variability, as captured by common measures of dispersion such as *variance/standard deviation* and *correlations* between its various features. Preserving such variability is used as the primary criterion to search for and assess methods to reduce dimensionality.

Example 3.2 (Principal Component Analysis (PCA)) Principal Component Analysis (PCA) is clearly the most popular DR method. When the dimensionality p is large, PCA is commonly used in exploratory data analysis and for extracting features and developing models. The key concept here is *principal component* (PC) of variability. The first principal component is the direction that maximizes the variance of a projection of the data onto a single line in feature space. The second principal component can be taken as a direction orthogonal to the first principal component that maximizes the variance of the remaining components of the projected data, as illustrated in Fig. 3.1.

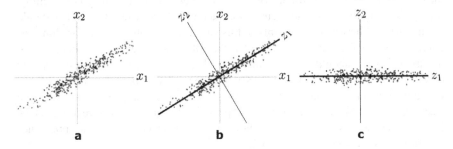

Fig. 3.1 (**a**) The first principal component (z_1) of a 2D dataset is best captured by (**b**) a projection along the solid black line (z_1) and its orthogonal second principal component (z_2). (**c**) plot of the recoding of the data in the new coordinate system with axes the principal components z_1 and z_2

Iterating the process, PCA computes the third principal component orthogonally to the first two PC's. PCA reduces dimensionality by projecting each data point onto only the first few principal components to obtain lower dimensional data representations, while preserving as much of the data's variation as possible. Overall, PCS amounts to a change of basis of the coordinate axes in feature space (e.g., by appropriate rotations, as illustrated in Fig. 3.1) in order to capture the most variability of the data along the new axes. Thus, PCA extracts features that retain the most amount of the variance/covariance in the high dimension data. Since these projections are linear operators, they can be easily implemented using a singular value decomposition (SVD) of its variance–covariance matrix with optimized and very fast linear algebra software libraries.   □

Example 3.3 (Kernel PCA (KPCA)) KPCA is an extension of PCA obtained by choosing a transformation of the data by a so-called *kernel* that defines some "weighted distance" measure. PCA is recovered as a particular case when the kernel is linear. Other kernels can be considered, such as (a) polynomial kernels, (b) Gaussian kernels, and (c) Laplacian kernels.   □

Example 3.4 (Independent Component Analysis (ICA)) Both PCA and ICA share the common feature of finding a set of vectors as a basis to re-code the data. PCA can greatly compress the data into fewer dimensions. On the other hand, ICA is useful to find a representation of high-dimensional data as independent subelements that can be used to separate data. Therefore, ICA is useful when the data (usually image data) is a mixture of multiple signals for the separation of the various independent components. ICA is used mostly for the purpose image processing.   □

3.2.2 Geometric Approaches

From a geometric perspective, PCA can be viewed as fitting a linear (flat) subspace to the data so as to minimize the error given by the total sum of squared distances of the data points from the subspace of reduced dimension. In a more general setting, the data may not be "flat," so a curved surface may best fit the data. What should be the distance between data points in this case (the straight distance or along the curve)? Also, while the dimension of a flat affine subspace is well defined and understood, what should be the dimension of a curved surface? The appropriate concept to better discuss these issues is the concept and language of manifolds. Roughly speaking, a manifold is a geometric object that appears to be flat like a Euclidean object (line, plane, affine hyperplane), for example, a sphere as big as the Earth. Some variants of this idea for DR are briefly sketched next. These approaches will be discussed in detail in Sect. 5.2.

The general idea of a geometric approach is to reduce dimensionality while minimizing some appropriate loss function that measures discrepancies in the distances, or some other function of the distances is minimized. The notion of distance being used defines a specific method.

Example 3.5 (Multi-Dimensional Scaling (MDS)) MDS has its origins in the field of psychometrics and it can be used to visualize the level of similarity of individual cases of a dataset by translating information about the pairwise "distances" among a set of *n* objects or individuals into a configuration of points mapped into an abstract Euclidean space. It can be viewed as a form of nonlinear dimensionality reduction from a given distance matrix. When the distance matrix actually consists of Euclidean distances, a procedure based on Linear algebra can be used to find a mapping of the objects to points in a Euclidean space, known as *classical MDS*. The algorithm will succeed and is in fact equivalent to the PCA algorithm, if the distances are actually Euclidean.   □

Example 3.6 (Isometric Mapping (ISOMAP)) ISOMAP is a mapping that preserves a distance measure defined over a lower dimensional manifold that can fit the data reasonably well. Its basic philosophy is to assume that "learning" this manifold is key to successful data analysis, and that the distances between the points are the "intrinsic" or geodesic distances between the points on the manifold. The key issues are: (a) how to build algebraic equations for a manifold from the sampled data; and (b) the effect of such a mapping into a lower dimensional space on the dataset for problem solving.   □

Example 3.7 (t-Stochastic Neighbor Embedding (t-SNE)) *t*-SNE is considered the state of the art in visualization algorithms. The *t*-SNE method, proposed by van der Maaten and Hinton [1], is a complex method that builds over and addresses the shortcomings of its precursor the *stochastic neighbor embedding* (SNE) method by Hinton and Roweis [2]. SNE is a probabilistic approach transforming objects given by high-dimensional vectors or by pairwise dissimilarities, into a lower dimensional space in such a way that neighbor identities are preserved. Unlike other dimensionality reduction methods, SNE can represent each object with a mixture of widely separated lower dimensional images.   □

In summary, MDS can preserve dissimilarities between items, as measured either by Euclidean distance, some nonlinear squashing of distances, or shortest graph path lengths as with ISOMAP. Principal components analysis (PCA) finds a linear projection of the original data which captures as much variance as possible. SNE can place the objects in a lower dimensional space so as to optimally preserve neighborhood identity and can be naturally extended to allow multiple different low dimensional images of each object. The *t*-SNE method is an improvement on the SNE method. A detailed discussion and comparison are given in Chap. 5.

3.2.3 Information-Theoretic Approaches

One can also perform a dimensionality reduction procedure using Shannon's concept of conditional entropy (CE). The basic idea is to select features/input variables by minimizing conditional entropy or, equivalently, to select features/input variables by maximizing mutual information. Specifically, mutual information measures the reduction in uncertainty for one variable when the value of another variable is known. This concept is closely related to the concept of information gain in machine learning, as a quantitative measure of the reduction in entropy, given the values of another random variable. The reduction can be used either to select predictor features using conditional entropies on others or to select them supervised mode using conditional entropies with the response/target variable.

Another interesting variation is to select features based on a recursive procedure by selecting the first feature as before to maximize the average single conditional entropy and make further selections by conditioning on previous choices. Unlike other DR methods, CE reduction can be used in both modalities, supervised and unsupervised learning. A detailed discussion is given in Chap. 6.

3.2.4 Molecular Computing Approaches

The key motivation of this approach is to exploit structural properties built deeply into DNA by millions of years of evolution that can be utilized for extreme dimensionality reduction and solution efficiency. They can be naturally used with genomic data, but perhaps surprisingly, with ordinary abiotic data just as well. There are several variations of this idea, and two major representative families of techniques of this kind are reviewed in detail in Chap. 7, namely genomic and pmeric coordinate systems for DNA sequences below.

Example 3.8 (Reduction by Genomic Signatures) It is based on the pointwise hybridization patterns exhibited by encodings of the data into DNA sequences to a common judiciously selected set of DNA oligonucleotides (a noncrosshybridizing DNA microarray or chip) of the same length blanketing the entire DNA space, and hence the dataset. □

Example 3.9 (Reduction by Pmeric Signatures) Another way of reducing dimensionality of genomic sequences is to capture their hybridization affinity (in terms of their hybridization distance (the h-distance), as described in Chap. 7), to some carefully selected oligonucleotides (the centroids of DNA spaces of a fixed length) that have the capacity to represent the whole DNA space and DNA sequences of arbitrary length. The method is based on the barycentric coordinates of the data points (after translation into DNA spaces) in the convex hull of the Euclidean representations of the centroids of DNA spaces of a fixed length. □

Similar to the categorization of supervised/unsupervised machine learning, dimensionality reduction methodologies can be categorized by whether the information of the response/target feature is used. Based on the specific target problem or research goal, an objective function can be identified to find an optimal solution to the problem. Specifically, dimensionality reduction can be achieved by optimizing an objective function. For example, to obtain the optimal linear combination to retain the maximal variation, the objective function will be the variance of a candidate linear combination. PCA is then the algorithm that provides the optimal solution.

3.3 The Blessings of Dimensionality

How does the dimensionality of the data impact data analyses? To answer this question, one can look at the geometric properties of high-dimensional spaces that are relevant for various computational aspects of data analysis. It turns out that high dimensionality can be both beneficial as well as problematic. The first of these facts is referred to as the *blessing of dimensionality* and it is somewhat lesser known than its counterpart, the *curse of dimensionality*. They are both sides of the same coin, as they depend on the same bare mathematical facts. This section describes the blessings of high dimensionality, while the curse is deferred to Sect. 10.6.

Both the blessing and the curse of dimensionality can be traced back to the same fundamental reason. In a nutshell, *there is too much room in higher dimensional spaces.* There is no pun intended here because the meaning of this multifaceted statement can be comprehended by doing some calculations about volume (the quantification of space) in higher dimensions and following up on their implications for DR and problem solving.

Example 3.10 On the 1D line, any point x separates other points in the 1D line in two sides so that it is impossible to pass from one to the other without passing through the point x. However, some sets of two points $S = \{x, y\}$ cannot always be separated from a third point z, for example, if z lies in between x and y. There is *not enough room* to accommodate or fit a separator between them (S) and z. However, in 2D (or 3D) space, unless x, y, z are not in general position (i.e., are collinear or coplanar) as above, there are infinitely many 1D lines (2D planes, respectively) that will separate the point z from the set S, i.e., put it on opposite sides of the flat separator.   □

Another subtle way in which high dimensionality manifests itself, but still a consequence of the fundamental reason above, is that "in higher dimensions, mass tends to be concentrated near the boundary." This is a fact known in probability theory as the *concentration of measure phenomenon* [3]. Several consequences that were first termed as blessings of dimensionality by Donoho [4] depend on it. They are referred to as blessings [4] since they help computationally. For example, they have been very useful in the design of randomized algorithms and data structures [5].

In particular, one of the consequences is that a so-called *Lipschitz function* (i.e., a function whose values $f(x)$ change slowly from the value $f(x_0)$ with the distance $|x - x_0|$ from x_0) can be estimated with high probability in high dimensions by sampling it at single points randomly. Another consequence is that nearest neighbor search in high dimensions can be approximately solved very quickly with high probability.

The nearest neighbor problem [**NNbrs**] for a given dataset $\mathbf{D} \subset \mathbf{R}^d$ of dimensionality d was stated precisely in Sect. 2.2.1 as follows.

[**NNbrs**]   **NEAREST NEIGHBOR (D)**

> INSTANCE: A point $x \in \mathbf{R}^d$
> QUESTION: Which is the nearest point in $\mathbf{D}$ to point x?

Example 3.11 Such queries are useful in Geographic Information Systems (such as your favorite Map app), for example, to answer queries like "Where is the nearest gas station?" (i.e., $\mathbf{D}$ is the set of gas station locations and x is your current location.) Such nearest neighbor queries happen often and, therefore, it is desirable to answer such queries fast.   □

Solutions to the nearest neighbors and the k nearest neighbors (**kNN** s) problems are important algorithmic constructs in higher dimensions, particularly for the purposes of classification, clustering, and function approximation (described in more detail in Sect. 2.2.1). A linear scan of all n points in $\mathbf{D}$ will yield the closest point in linear time, but are there substantially faster ways to answer such queries? Indeed, in low dimensions such as 2D, one can preprocess the points into a data structure (a Voronoi diagram) of size in the order of n (i.e., $O(n)$) such that nearest neighbor queries can be answered in $O(\log n)$ time. In higher dimensions, however, although one can still precompute such a Voronoi diagram, the curse of dimensionality rears its ugly head as a *combinatorial explosion*. In higher dimensions, a Voronoi diagram becomes an exploding collection of points, edges, 2D faces, 3D and higher dimensional polytopes (about $O(n^{d/2})$ of them), which are computationally infeasible to store or search (more later in Sect. 10.6). The curse of dimensionality strikes again!

On the other hand, there is a blessing too that helps alleviate this situation, if one is willing to accept only approximate solutions. In the *approximate nearest neighbor search* one usually has some stringency parameter $\varepsilon > 0$ and instead of asking for the exact nearest neighbor, one settles with an $(1+\varepsilon)$-approximate nearest neighbor, i.e., a point z' in $\mathbf{D}$ whose distance $|z'-x|$ from the query point is within $(1+\varepsilon)$ times the distance to the true nearest neighbor z. The hashing scheme is called *locality sensitive hashing* and is a well-known acceptable solution to this problem. Such algorithms have been very successful and they afford sublinear query time (great savings when looking for a gas station) compared to a linear scan. This guarantee can be given for this solution design only because of the concentration of measure phenomenon mentioned above.

There are yet other good consequences of the blessing of dimensionality, as pointed out by Donoho and Tanner [6], and Barany and Furedi [7]. Roughly speaking, they show that if we sample data points independently in higher dimensions with, say, a standard normal distribution (or uniform within a ball), it is true that, with high probability, each point is a vertex of the convex hull of all the points, even for a sample size exponentially large in the dimension d. Intuitively, this is true since all points sampled will lie near the boundary with high probability, as mentioned above. In particular, this means that *each point can be separated from all the others by a hyperplane*. This fact is important in the design of correctors in AI systems [8], for example. More specifically, if an AI system makes a mistake, some decision rules need to be changed to address it. However, since a complete re-training of the system may be infeasible on every mistake, a re-training is postponed until several mistakes are encountered. Using the linear separability mentioned above, simple correctors can be designed that can cause an AI system to change its decision on a single point to avoid future mistakes without affecting other points, unless the number of mistakes is too large.

References

1. Maaten, L., & Hinton G. E. (2008). Visualizing data using t-sne. *Journal of Machine Learning Research, 9*(86), 2579–2605.
2. Hinton, G. E., & Roweis, S. T. (2002). Stochastic neighbor embedding. In *Advances in Neural Information Processing Systems* (Vol. 15, pp. 833–840). The MIT Press.
3. Ledoux, M. (2001). *The Concentration of Measure Phenomenon.* American Mathematical Society.
4. Donoho, D. L. (2000). High-dimensional data analysis: The curses and blessings of dimensionality. In *AMS Conference on Math Challenges of the 21st Century.*
5. Dubhashi, D. P., & Panconesi, A. (2009). *Concentration of measure for the analysis of randomized algorithms.* Cambridge University Press.
6. Donoho, D. L., & Tanner, J. (2009). Observed universality of phase transitions in high-dimensional geometry with implications for modern data analysis and signal processing. *Philosophical Transactions of the Royal Society A: Mathematical, Physical and Engineering Sciences, 367,* 4273–4293.
7. Barany, I., & Füredi, Z. (1988). On the shape of the convex hull of random points. *Probability Theory and Related Fields, 77,* 231–240 (1988)
8. Tyukin, I. Y., Gorban, A. N., McEwan, A. A., Meshkinfamfard, S., & Tang, L. (2021). Blessing of dimensionality at the edge and geometry of few-shot learning. *Information Sciences, 564,* 124–143 (2021)

Chapter 4
Conventional Statistical Approaches

Ching-Chi Yang (iD), **Max Garzon** (iD), **and Lih-Yuan Deng** (iD)

Abstract The objective of dimensionality reduction is to retain key properties of the given data to solve a problem with fewer features in a lower dimensional space. Statistical methods aim to preserve characteristic parameters such as mean, variance, and covariance of features in the population, as estimated from the dataset. Methods include Principal Component Analysis (PCA) and its variants, Independent component analysis and Discriminant Analysis. Linear algebra methods offer other approaches, including Singular value Decomposition (SVD) and Nonnegative Matrix Factorization (NMF).

4.1 Principal Component Analysis (PCA)

Principal component analysis (PCA) extracts features that retain the most amount of the variance/covariance of a dataset and can be easily implemented using elementary methods in statistics and linear algebra (summarized in Sects. 11.1 and 11.2). The variance is the measurement of how spread out the data is in the set, a measure of variability. PCA takes a dataset with a lot of dimensions and flattens it to fewer dimensions (say 2D or 3D) so the power of the human eye can look at it and get a deeper understanding of the data. It is also a classical feature extraction and data representation technique widely used in the areas of pattern recognition and computer vision such as face recognition. Therefore, PCA is a classical and very popular dimensionality reduction method.

Example 4.1 (PCA with Iris Data) The Iris dataset was introduced in Sect. 1.1 (also described in Sect. 11.4). In the problem **[IrisC]** of classifying iris flowers, one

C.-C. Yang (✉) · L.-Y. Deng
Mathematical Sciences, The University of Memphis, Memphis, TN, USA
e-mail: cyang3@memphis.edu; lihdeng@memphis.edu

M. Garzon
Computer Science, The University of Memphis, Memphis, TN, USA
e-mail: mgarzon@memphis.edu

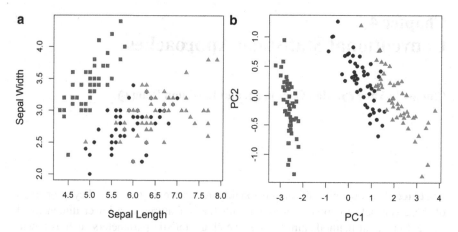

Fig. 4.1 (**a**) Scatter plot based on Sepal Length and Sepal Width. (**b**) Scatter plot based on the first and the second principal components. The grey levels of the points represent three different types of species

class is linearly separable (by a hyperplane) from the other two, but the latter are *not* linearly separable from each other. As expected, the measurements in the four features (SepalLength, SepalWidth, PetalLength, and PetalWidth) are correlated. PCA extracts features that will be orthogonal to the others. The extracted features, then, will be treated as uncorrelated features, so they might confuse a classifier less, while at the same time attempting to retain the most variation in the data to solve the problem. As mentioned in Sect. 3.2, the key concept in PCA is *principal component* onto which the data is being projected to find these components. Figure 4.1b shows the result after projecting the Iris dataset onto the first two principal components.   □

4.1.1 Obtaining the Principal Components

PCA projects the original data in a direction that maximizes variance, as illustrated in Fig. 3.1 in Sect. 3.2 and Fig. 4.1b. The first principal component (PC_1) is the axis that affords the most variation when projecting the data onto a single 1D axis.

In general, the dataset $\mathbf{X}$ can be first *centered* so that their sample mean is 0 (by subtracting the mean of the feature values from each feature X_i). The first component PC_1 is chosen as a column vector $\mathbf{W}_1$ satisfying

$$\mathbf{W}_1 = \arg \max_{||w||=1} \{||\mathbf{XW}||^2\} = \arg \max_{||w||=1} \{\mathbf{W}'\mathbf{X}'\mathbf{XW}\}. \tag{4.1}$$

Thereafter, the kth principal component PC_k ($k > 1$) is found recursively as the axis that spans the most variation in residual projections of the data onto directions orthogonal to the first component. The kth greatest variance lies along the kth component. PC_k is aimed to obtain a column vector $\mathbf{W}_k$ such that

$$\mathbf{W}_k = \arg \max_{\|\mathbf{W}\|=1} \{\|\hat{\mathbf{X}}_k \mathbf{W}\|^2\} = \arg \max_{\|\mathbf{W}\|=1} \{\mathbf{W}'\hat{\mathbf{X}}_k'\hat{\mathbf{X}}_k\mathbf{W}\}, \tag{4.2}$$

where

$$\hat{\mathbf{X}}_k = \mathbf{X} - \sum_{i=1}^{k-1} \mathbf{X}\mathbf{W}_i\mathbf{W}_i'.$$

Therefore, PCA can be regarded as an orthogonal (i.e., perpendicular axes) linear transformation that recodes data into a new coordinate system so that the greatest variance by some projection of the data lies on the axes of the new system, i.e., the principal components. However, directly solving the PCA optimization problem might not be efficient. Section 4.1.2 presents another approach to solving the PCA optimization problem via linear algebra.

PCA measures the variation based on Euclidean distances between the data points. Some researchers have extended the idea to other metrics. Other approaches and extensions are discussed in Sect. 4.2.

4.1.2 Singular Value Decomposition (SVD)

The variance optimization problem admits a very nice solution using standard linear algebraic techniques to compute eigenvectors (summarized in Sect. 11.2). The corresponding eigenvectors yield the principal components, as shown next.

Singular value decomposition (SVD) is a well-known matrix factorization method in linear algebra. It decomposes an $n \times p$ matrix $\mathbf{X}$ into a factorization of the form

$$X = \mathbf{UDV}',$$

where $\mathbf{U} = (\mathbf{U}_1, \mathbf{U}_2, \ldots, \mathbf{U}_n)$ is an $n \times n$ orthogonal matrix, $\mathbf{V} = (\mathbf{V}_1, \mathbf{V}_2, \ldots, \mathbf{V}_p)$ is a $p \times p$ orthogonal matrix, and $\mathbf{D}$ is an $n \times p$ rectangular matrix with nonzero elements along the first $p \times p$ submatrix diagonal with $n > p$, that is, $\mathbf{D} = \Lambda(d_1, d_2, \ldots, d_p)$ where $d_1 \geq d_2 \geq \cdots \geq d_p \geq 0$. (Typically, the p elements of the submatrix diagonal are organized from the largest value to the smallest one. A matrix $(\mathbf{U}_1, \mathbf{U}_2, \ldots, \mathbf{U}_n)$ is called *orthogonal* if its column vectors $\mathbf{U}_i$ and $\mathbf{U}_j$ are orthogonal, i.e., their inner product $\mathbf{U}_i'\mathbf{U}_j = 0$ for all pairs $i \neq j$ and $\mathbf{U}_i'\mathbf{U}_i = 1$ for all i. Thus, the inverse of an orthogonal matrix is its transpose, i.e., the product

$\mathbf{D'D} = \mathbf{I}$.) If one re-expresses the data matrix $\mathbf{X}$ by $\mathbf{UDV'}$, the variance of $\mathbf{XW}$ can be written as

$$\frac{\mathbf{W'X'XW}}{n-1} = \frac{\mathbf{W'VD'DV'W}}{n-1}.$$

It can be shown that one will obtain the largest variance $d_1^2/(n-1)$ by making $\mathbf{W} = \mathbf{V}_1$, with value

$$\mathbf{W'X'XW} = \mathbf{V}_1'\mathbf{X'XV}_1 = \mathbf{V}_1'\mathbf{VD'DV'V}_1 = \mathbf{e}_1'\mathbf{D'De}_1 = d_1^2,$$

where $\mathbf{e}_1$ is the first base vector of dimension p. Similarly, if one uses $\mathbf{V}_k$ instead of $\mathbf{V}_1$, the linear combination $\mathbf{XV}_k$ will result in the kth largest residual variance. The column vectors of $\mathbf{V}$ are the *eigenvectors* of $\mathbf{X'X}$ and the values d_i^2 are the square of the *eigenvalues* of $\mathbf{X'X}$.

Example 4.2 (SVD for the Iris Dataset) If SVD is implemented on the centered 150×4 matrix Iris data $\mathbf{X}$ (by subtracting the mean value from each features), the matrices $\mathbf{U}$, $\mathbf{D}$, and $\mathbf{V}$ will be a 150×150 orthogonal matrix, a 150×4 matrix with 4 diagonal nonzero elements, and a 4×4 orthogonal matrix, respectively. Since $\mathbf{U}$ will contain all zero entries from the 5th to the 150th column, the matrix can be simplified to a 150×4 matrix. Similarly, the matrix $\mathbf{D}$ will be simplified to a 4×4 diagonal matrix.

$$\mathbf{U} = \begin{bmatrix} -0.11 & -0.05 & 0.01 & 0.00 \\ -0.11 & 0.03 & 0.06 & 0.05 \\ -0.12 & 0.02 & -0.01 & 0.01 \\ & \cdots & & \\ 0.06 & 0.05 & -0.11 & -0.08 \end{bmatrix}, \quad \mathbf{D} = \begin{bmatrix} 25.10 & 0.00 & 0.00 & 0.00 \\ 0.00 & 6.01 & 0.00 & 0.00 \\ 0.00 & 0.00 & 3.41 & 0.00 \\ 0.00 & 0.00 & 0.00 & 1.88 \end{bmatrix} \quad \text{and}$$

$$\mathbf{V} = \begin{bmatrix} 0.36 & -0.66 & 0.58 & 0.32 \\ -0.08 & -0.73 & -0.60 & -0.32 \\ 0.86 & 0.17 & -0.08 & -0.48 \\ 0.36 & 0.08 & -0.55 & 0.75 \end{bmatrix}$$

□

4.2 Nonlinear PCA

So far, PCA has been considered only for populations and data living in Euclidean spaces with Euclidean distances. Chapter 5 will pursue refinements of PCA in other spaces. As the Manhattan distance shows, other distances might be more suitable for other datasets and populations. This section discusses some techniques extending the idea to get some variations of PCA for data with other metrics in other spaces (although not to metric spaces of infinite dimension, such as Hilbert spaces). *Kernel PCA* (KPCA) maximizes variance along *other bases*, while *Independent Component Analysis* (ICA) focuses on the *independence* of the features (the PC axes).

4.2.1 Kernel PCA

Two major strengths of PCA are that one obtains the maximal variation with a small set of features along PCs and the PCs are orthogonal and hence linearly independent. This section explores the case where the data can be transformed into other metric spaces. In a different space, the variation may have to be measured differently. PCA-like algorithms in such spaces are generally called *kernel PCA*.

Example 4.3 (Kernel PCA with Iris Data) A change in distance metric can cause PCs to look very different, even for the same dataset. The first two princi-pal components of the Iris data set with different kernel metrics are shown in Fig. 4.2. □

Section 4.1 showed how PCA aims to maximize the value of the distances $\|\mathbf{XW}\|^2$. They can be rewritten as an inner product $\mathbf{XW \cdot XW}$ which is just $\mathbf{W'X'XW}$. As described in Sect. 2.2, SVMs can change the inner product to better describe the distance between observations to facilitate solving a problem. The recoding function $K(\mathbf{x}_i, \mathbf{x}_j)$ is called a *kernel function*. It re-scales the data points to make analysis by ordinary linear PCA possible. In general, the distances can be weighted. By utilizing different weight functions $K(\mathbf{x}_i, \mathbf{x}_j)$, the separation boundary will fit the data better and improve performance. Commonly used kernels are

- *Linear kernel* (no recoding)
  Essentially PCA itself using Euclidean distance.
- *Polynomial kernel*
  $K(\mathbf{x}_i, \mathbf{x}_j) = (\mathbf{x}_i\mathbf{x}_j' + 1)^d$, recoding by a polynomial of degree d. The distance between two observations will be enlarged by a factor of a power d of the distances. (The formula contains a shift "+1" to guarantee that the distance is larger when $\mathbf{x}_i' \cdot \mathbf{x}_j'$ is less than 1.)
- *Gaussian kernel*
  $K(\mathbf{x}_i, \mathbf{x}_j) = \exp\{-\|\mathbf{x}_i - \mathbf{x}_j\|^2/2\sigma^2\}$. Instead of enlarging the distance between two observations, one can instead pull the "outliers" back in. A Gaussian distributions can be used, such as $f(x) \propto \exp\{-x^2/\sigma^2\}$. The "distance" should rather be interpreted as similarity now because the metric properties from Sect. 1.2 may fail. If $\|\mathbf{x}_i - \mathbf{x}_j\|^2$ is larger, the value of $K(\mathbf{x}_i, \mathbf{x}_j)$ is near 0 (no similarity); if $\|\mathbf{x}_i - \mathbf{x}_j\|^2$ is 0, the value of $K(\mathbf{x}_i, \mathbf{x}_j)$ is 1 (no change in the distance).

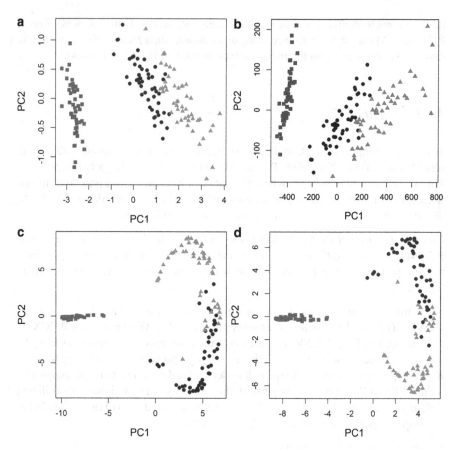

Fig. 4.2 Scatter plots of the first two principal components (PCs) of the Iris dataset with nonEuclidean distances based on (**a**) linear kernel; (**b**) polynomial kernel with two degrees of freedom; (**c**) Gaussian kernel with $\alpha = 1$; (**d**) Laplacian kernel with $\alpha = 1$

- *Laplacian/exponential kernel*

  $K(\mathbf{x}_i, \mathbf{x}_j) = \exp\{-\|\mathbf{x}_i - \mathbf{x}_j\|/\sigma\}$. Instead of using $\|\mathbf{x}_i - \mathbf{x}_j\|^2$ in determining the similarity, one can also consider using $\|\mathbf{x}_i - \mathbf{x}_j\|$ by leaving out the square of the norm.

Generally, Gaussian and Laplace kernels are called *radial basis kernels*.

4.2.2 Independent Component Analysis (ICA)

ICA is a computational technique for decomposing a multivariate signal into additive subcomponents. The objective now is to obtain "independent" components $\mathbf{u}_1, \mathbf{u}_2, \ldots, \mathbf{u}_k$ of dimension p, so that $\mathbf{x}_i = w_{i_1}\mathbf{u}_1 + w_{i_2}\mathbf{u}_2 + \cdots + w_{i_k}\mathbf{u}_k + \varepsilon$, where ε is a Gaussian error of dimension p.

Example 4.4 (ICA with Iris Data) Although ICA is commonly used in image recognition, it is still possible to utilize it on the Iris dataset. (to be fair, one should randomize the order of the observations.) One can obtain the independent "signals" based on ICA. Figure 4.3a shows the first two independent components (ICs) and Fig. 4.3b shows the ICs with respect to their observation indices. The first independent components in Fig. 4.3a already separate Iris species well.

Similar to decorrelation of features in PCA by principal components in Sect. 4.1, a *pre-whitening* procedure decorrelates each observation data matrix by its mean by performing an eigenvalue decomposition on $\mathbf{XX}' = \mathbf{UDD}'\mathbf{U}'$. The decorrelated observations will simply be $\mathbf{DU}'\mathbf{X}$. So, without loss of generality, one can assume that $\mathbf{X}$ is pre-whitened data. The objective of ICA is then to find the direction for the weight vector $\mathbf{W}$, a column vector of dimension n, that maximizes either nonGaussianity or mutual information. This section focuses on nonGaussianity. Mutual information-related topics will be described in Chap. 6.

FastICA is an efficient and popular algorithm for ICA invented by Aapo Hyvärinen at the Helsinki University of Technology. Like most ICA algorithms, FastICA seeks an orthogonal rotation of pre-whitened data through a fixed-point iteration scheme that maximizes a measure of nonGaussianity of the rotating

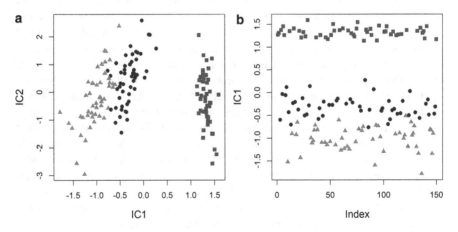

Fig. 4.3 (a) Scatter plot of the first two independent components of the Iris dataset. (b) The observations are fairly separated in the first independent component and can be used to solve the problem of [**IrisC**]

components. Hyvärinen [6] suggests the nonlinear function $f(u) = \log(cosh(u))$ for general purposes and $f(u) = -\exp\{-u^2/2\}$ for robustness, as described next.

Let $g(x) = f'(x)$ and $h(x) = f''(x)$ denote the first and the second derivative of the function f. The FastICA algorithm obtains first component as follows:

1. Initialize the weight vector $\mathbf{W}$ randomly;
2. Update expectations and weights using

   - Expectation: $\mathbf{v} = \mathbf{X}(g(\mathbf{W'X}))' - (h(\mathbf{W'X})\mathbf{1}_p)\mathbf{W}$;
   - Weights: $\mathbf{W} = \mathbf{v}/\|\mathbf{v}\|$;

3. Iterate steps 1 and 2 until $\mathbf{W}$ has converged.

Thus, the first independent component s_1 is $\mathbf{W'X}$ and the final weight is denoted by $\mathbf{W}_1$. To obtain the kth independent component, a similar procedure is performed, with the additional constraint that it should be orthogonal to the previously obtained ICs in the second step:

- Expectation: $\mathbf{t} = \mathbf{X}(g(\mathbf{W'X}))' - (h(\mathbf{W'X})\mathbf{1}_p)\mathbf{W}$,
- Orthogonalization: $\mathbf{v} = \mathbf{t} - \sum_{j=1}^{k-1}(\mathbf{t'W}_j)\mathbf{W}_j$;
- Weights: $\mathbf{W} = \mathbf{v}/\|\mathbf{v}\|$.

Once $\mathbf{W}$ has converged, one obtains the kth independent component s_k as $\mathbf{W'X}$ and the final weights as $\mathbf{W}_k$.

4.3   Nonnegative Matrix Factorization (NMF)

In this section, a further exploration of linear algebra methods along matrix factorization for DR is shown to be useful for certain kinds of problems involving semantic concepts (i.e., abstract ideas such as *word meaning*, usually expressed as images or text). The key idea here is that feature *abstraction* as a combination of raw features is better than just feature selection as a way to produce fewer informative features for solving a DS problem. The key difficulty is to find the right criterion to combine features effectively. PCA is a very good example based on the criterion of capturing the variance/covariance of the sample by projections onto individual axes (the principal components). Formally, it amounts to an SVD decomposition of the data $\mathbf{X}$ into a product of three matrices $\mathbf{X} = \mathbf{UDV'}$, where $\mathbf{D}$ is a (possibly rectangular) diagonal matrix of eigenvalues. In Sect. 4.2 the idea was focused on different combinations of axes (linear bases) with kernel PCA and on the independence of the components. In this section, focus is shifted rather to more semantic and abstract features that may extract deeper structural properties of the data, where negative values can make intuitive interpretations difficult, as illustrated in the following examples.

Example 4.5 An interesting problem in understanding abstract concepts is to capture commonalities among a social group of n individuals (say, users), for example, in terms of liking or disliking specific p items (say, movies), where n and p are fairly large. A good recommender system for movies would have to "understand" why exactly it is that people tend to like certain movies but dislike others. A specific individual may give a high rating to a movie simply because her favorite actor is the main character. Another may give it a low rating because the topic is LGBT. If the data consists of a matrix $\mathbf{X}_{n \times p}$ of ratings x_{ij} of users i of movies j, it would be desirable to identify a few features (say, main characters, topics, directors, whether it is an academy award nominee, and so forth) that primarily determine the rating for an individual user. These features are considered *latent* or hidden, in the sense that they are not mentioned explicitly in data or ratings, but weigh heavily on them. Each such feature can be thought of as adding a bit of a "like" component to the whole group rating of a given movie, so that each column $\mathbf{X}_j$ of $\mathbf{X}$ expressing the group's sentiment for a specific movie j is actually a linear (weighted) combination of little "likes" contributed by each user for each latent feature, i.e.,

$$\mathbf{X}_j = \mathbf{U}\mathbf{V}_j,$$

where $\mathbf{V}_j$ is the jth column of a certain matrix $\mathbf{V}$ capturing all those little "likes" (which must be nonnegative), and $\mathbf{U}$ is the recoding of the features in $\mathbf{V}$ to retrieve the data $\mathbf{X}$. Mathematically, this simply means that the matrix $\mathbf{X}$ is being decomposed into a *product* $\mathbf{X} = \mathbf{U}\mathbf{V}$ of two (perhaps nonsquare rectangular) matrices $\mathbf{U}$ and $\mathbf{V}$ with nonnegative entries, a so-called *nonnegative matrix factorization* (NMF). Finding such a decomposition will deliver a recommender system for the problem of predicting the rating of a random user would give to a particular movie, or even for finding a movie to recommend to a particular user (by finding the movies with such top ratings).   □

Actually, the r rows of $\mathbf{V}$ will correspond to the latent features to be selected, while the number of columns is p, the number of movies. An algorithm solving the NMF problem generates these abstract features, so they do not have to correspond to features that would necessarily be intuitively understandable to humans. Although thus less explainable, they have the advantage of being objective (observer independent) since they are up to the mathematics of matrix multiplication.

Example 4.6 (Latent Semantic Analysis) Another challenging problem in Natural Language Processing (NLP) is to formulate the concept of *word meaning* in an objective and user-independent way, so that a machine could have some sense of the semantics of a word in a natural language like English and use it to communicate better with humans. The problem can be solved by an SVD factorization of the matrix $\mathbf{X}$ counting the number of occurrences of each of the p words in n documents, of the form

$$\mathbf{X} = \mathbf{U}\mathbf{D}\mathbf{V}',$$

where **D** is a rectangular diagonal matrix and **U**, **V** perform the recoding of the original data from **D**. The factorization altogether would determine the meaning of a word through the various uses of the word in the documents, as discussed in Sect. 4.2. (Note that the matrix factors may have negative entries. Section 11.2 describes SVD decomposition in some detail.) The feature vector coding the words and texts correspond to projections of the corresponding original frequency representations onto the eigenvectors of a suitable selection of the top eigenvalues in **D**.   □

Example 4.7 (k-Means Clustering by Factorization) The third problem where factorization can help is a clustering problem, specifically the k-Means algorithm discussed above in Sect. 2.1. As discussed in Sect. 4.2, the problem can be solved by a similar NMF factorization of the matrix

$$\mathbf{X} = \mathbf{U}\mathbf{D}\mathbf{V}'$$

counting the number of occurrences of p words in n documents, which together would determine the meaning a word through the various uses of the word in the sample of users.   □

Unlike with ordinary integers and prime numbers, a NMF of a matrix is not unique (i.e., there can be several equally useful but distinct concepts for an abstract feature) and can admit a large number of solutions; moreover, finding exact solutions is not only computationally hard (even with restrictions on the type of matrix **X**, e.g., symmetric with a diagonal submatrix of equal rank), and even if not all solutions are to be found. Additional criteria are needed to guide the search for the latent features. The problem of NMF can then be defined precisely as *an optimization problem where the factors only provide the search space for a matrix of nonnegative entries*, while the goal is to optimize the given criterion (a loss function or a cost function). For example, in PCA, the SVD decomposition aims at maximizing the variance of the data along orthogonal principal components in relation to the variance of the given matrix **X**. The problem can be further complicated where the data available is very large (e.g., a large number of users and movies, as was the case in the Netflix challenge problem (https://en.wikipedia.org/wiki/Netflix_Prize), incomplete, and/or sometimes even dynamically changing at high velocity (more users and new movies being added, e.g., in the problem of collaborative filtering in recommender systems).

4.3.1 Approximate Solutions

A solution to an NMF problem for a data set $\mathbf{X}_{n \times p}$ provides DR for data if the number of latent factors (intermediate dimension) is substantially smaller than both n and p. The explainability of the solution also requires the entries to be nonnegative. This constraint precludes the use of a number of other linear algebra techniques such as SVD to solve the problem.

One of the most popular solutions, because of the simplicity and ease of its implementation, is due to Lee and Sung [7]. The trade-off is that the solution is only approximate, i.e., the product of the factor matrices is only close to $\mathbf{X}$ according to some distance function in matrix spaces, such as the ℓ_2 function (ordinary distance in Euclidean spaces of dimension n^2). The strategy for this solution is the *multiplicative update rule* used in binary decision making based on the advice from various experts, whose opinions can be weighed (un)equally. A decision maker can simply make the first decision based on the majority of the experts' advice, weighed equally at first. In successive rounds, the weight of an advisor's opinion can be repeatedly updated based on the correctness of their prior advice. For NMF, the cost function can be chosen as the square of the ℓ_2-distance

$$\|\mathbf{X} - \mathbf{UV}\|_2^2 = \sum_{i,j} [\ \mathbf{X}_{ij} - (\mathbf{UV})_{ij}\ ]^2 \tag{4.3}$$

of the product of the estimated factors from the target $\mathbf{X}$. Given approximate factors $\mathbf{U}$ and $\mathbf{V}$ in the tth iteration, an *additive* update to improve $\mathbf{V}$ can be proportional to that difference in the $t + 1$st iteration, i.e.,

$$\mathbf{V}_{ij} \ \leftarrow \ \mathbf{V}_{ij} + \eta_{ij}[\ (\mathbf{U}^T \mathbf{X})_{ij} - (\mathbf{U}^T \mathbf{UV})_{ij}\ ],$$

and likewise for $\mathbf{U}$. The values of the learning rates η_{ij} can be obtained by gradient descent (described in Sect. 11.3) in matrix space, i.e., the projection of the components of the gradient (vector of partial derivatives) of the cost function (4.3) with respect to the entries in $(\mathbf{X} - \mathbf{UV})_{ij}$ along the (ij)th direction.

However, perhaps surprisingly, the *multiplicative update* rule with larger updates given entrywise by

$$\mathbf{V}_{ij} \ \leftarrow \ \mathbf{V}_{ij} \frac{(\mathbf{U}^T \mathbf{X})_{ij}}{(\mathbf{U}^T \mathbf{UV})_{ij}}$$

and

$$\mathbf{U}_{ij} \ \leftarrow \ \mathbf{U}_{ij} \frac{(\mathbf{XV}^T)_{ij}}{(\mathbf{UVV}^T)_{ij}}$$

will still converge to factors where the updates will be 0 at the local optimal values.

As a technical point, the monotonic convergence of the algorithm can be proven using an auxiliary function analogous to that used for proving convergence of the Expectation Maximization algorithm in Probability Theory [7] (described in Sect. 11.3.2). The algorithm can also be interpreted as a diagonally rescaled gradient descent, where the rescaling factor is optimally chosen to ensure convergence.

4.3.2 Clustering and Other Applications

With the least squares criterion, NMF is equivalent to a relaxed form of k-Means clustering. In k-Means clustering (described in Sect. 2.3) of data $\mathbf{X}$, if the cluster centroid vectors $\mathbf{c}_k$ are the columns of a matrix $\mathbf{H}$, then an optimal solution for hard clustering will minimize the aggregate inter-cluster distance between points across all clusters, i.e., the least squared error function

$$J_k(\mathbf{C}) = \sum_{j=1}^{k} \sum_{i \in C_k} \|\mathbf{x}_i - \mathbf{c}_j\|^2 , \qquad (4.4)$$

where $\mathbf{C}$ is the matrix of cluster centroids $\mathbf{c}_1, \ldots, \mathbf{c}_k$ and the C_ks are the clusters. In hard clustering, the *cluster indicator vector* for cluster $\mathbf{c}_i$ has a 1 or a 0 at position j depending on whether data point j belongs to the cluster $\mathbf{c}_i$ or not, so that indicator vectors for two different clusters are perpendicular vectors with dot product 0 because the clusters make a partition of the dataset (hence a 1 in one position forces 0 in the same position in all the other cluster indicator vectors). Thus, the matrix $\mathbf{H}$ with columns the indicator vectors in some ordering of the clusters is orthogonal, and the matrix $\mathbf{V} = \mathbf{HH}'$ can be used for a factorization. Therefore, when $\mathbf{U}$ is required to be orthogonal (i.e., $\mathbf{UU}' = \mathbf{I}$, the identity matrix) and $\mathbf{V} = \mathbf{U}'$ in a matrix factorization, a (an approximate) solution corresponds to an optimal (approximate, respectively) clustering of the data. The dominant entries $\mathbf{V}_{kj} > \mathbf{V}_{ij}$ determine a cluster C_k of the input data. This property tends to hold even if the orthogonality condition is not enforced. Furthermore, if the optimization criterion is the Kullback–Leibler divergence (defined precisely in Sect. 5.2), NMF is equivalent to the Latent Semantic Analysis popular in Natural Language Processing (NLP) for document clustering. (More details can be found in [2].)

NMF has a wide variety of applications in other fields, including the original application in astronomy (e.g., detecting exoplanets outside the solar system), computer vision (object identification), audio signal processing (nonstationary speech denoising), and bioinformatics (identifying gene cluster responsible for cancers by certain types of genetic mutations; predicting miRNA disease associations [3]; quantification of single-cell RNA gene expression [9]). A broad survey of NMF, its generalizations (tensor factorization) and their applications can be found in [1] and [5].

4.4 Discriminant Analysis

Fisher's Discriminant analysis [4] can be used to classify qualitative response variables into two or more classes. The method uses prior probabilities of belonging to each class together with density functions associated with predictor variables.

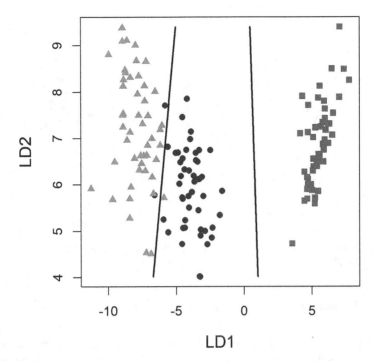

Fig. 4.4 Linear Discriminant Analysis (LDA) on the first two discriminant dimensions of the Iris dataset

Using Bayes theorem (described in Sect. 11.1), posterior distributions are calculated for the probability that an observed response belongs to each class given the value of the predictor variable. LDA and QDA are generalizations of Fisher's Discriminant Analysis. Linear discriminant analysis (LDA) assumes that observations within each class come from a multivariate normal distribution with different means for each class but a common covariance matrix for all classes. Quadratic discriminant analysis (QDA) makes a similar assumption but it allows each class to have its own covariance matrix.

Example 4.8 (LDA and QDA with Iris Data) The Iris data can be visualized by projecting the data into the first two LD dimensions. Figure 4.4 shows the decision boundaries based on LDA. By assuming the same covariance matrix, the boundaries will be straight lines, hence the name *linear* discriminant analysis.

The Iris dataset can be transformed to illustrate the advantages of LDA and QDA. Sharper boundaries between the classes can be established in both cases by just transforming Petal Length and Petal Width, as shown in Fig. 4.5. These boundaries will be linear if the covariance matrices for each class are similar. By allowing different covariance matrices for each class, the decision boundaries can be nonlinear. In fact, they are quadratic boundaries, hence *quadratic* discriminant analysis. □

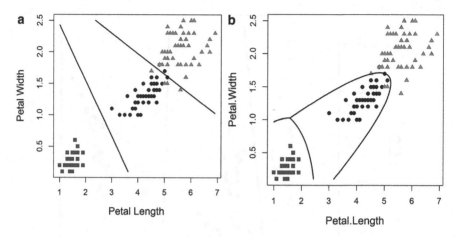

Fig. 4.5 Different decision boundaries on Petal Length and Petal Width for the Iris dataset via (**a**) LDA and (**b**) QDA. The classes are now more easily separable

4.4.1 Linear Discriminant Analysis (LDA)

Fisher's Linear discriminant analysis (LDA) aims to find a linear combination of the given variables that separates two or more classes. LDA uses Bayesian learning and Bayes' theorem (stated in Sect. 11.1). The main idea separating LDA and Bayesian learning is that LDA assumes that the distribution of the conditional probability $P(\mathbf{X} = \mathbf{x}|C = c)$ follows a Gaussian distribution with the same covariance matrix Σ. Specifically, the multivariate Gaussian distribution for $[\mathbf{X}|C = c]$ is given by

$$P(\mathbf{X}|C = c) = \frac{1}{(2\pi)^{p/2}|\Sigma|^{1/2}} \exp\left\{-\frac{1}{2}(\mathbf{x} - \mu_c)\Sigma^{-1}(\mathbf{x} - \mu_c)'\right\},$$

where μ_c is an estimated mean (a row vector of dimension p) for class value c and Σ is a $p \times p$ estimated covariance matrix. By Bayes theorem, $P(C = c|\mathbf{x}) \propto P(\mathbf{x}|C = c)P(C = c)$. Therefore, the decision boundary between class j and k is $\{\mathbf{x} : P(\mathbf{x}|C = j) P(C = j) = P(\mathbf{x}|C = k)P(C = k)\}$. By taking logarithms of both sides, the decision boundary will be determined by

$$\log(\pi_j) - \frac{1}{2}(\mathbf{x} - \mu_j)\Sigma^{-1}(\mathbf{x} - \mu_j)' = \log(\pi_k) - \frac{1}{2}(\mathbf{x} - \mu_k)\Sigma^{-1}(\mathbf{x} - \mu_k)',$$

which can be further simplified to

$$\log(\pi_j) - \frac{1}{2}\mu_j \Sigma^{-1}\mu_j' + \mathbf{x}\Sigma^{-1}\mu_j' = \log(\pi_k) - \frac{1}{2}\mu_k \Sigma^{-1}\mu_k' + \mathbf{x}\Sigma^{-1}\mu_k'.$$

Table 4.1 Prediction accuracies of LDA and QDA with Petal Length and Petal Width as input variables for the Iris classification problem [**IrisC**]

LDA	Setosa	Versicolor	Virginica	QDA	Setosa	Versicolor	Virginica
setosa (pred.)	50	0	0	setosa (pred.)	50	0	0
versicolor (pred.)	0	48	4	versicolor (pred.)	0	49	2
virginica (pred.)	0	2	46	virginica (pred.)	0	1	48

4.4.2 Quadratic Discriminant Analysis (QDA)

Instead of assuming the same covariance matrix across all classes, QDA assumes that the distribution of the random variable $[\mathbf{X}|\mathbf{C} = k]$ follows a Gaussian distribution with different covariances matrix Σ_k. Using a similar derivation to LDA in Sect. 4.4.1, the decision boundary for class j and class k will be given by

$$\log(\pi_j) - \frac{1}{2}\mu_j \Sigma_j^{-1}\mu_j' + \mathbf{x}\Sigma_j^{-1}\mu_j' = \log(\pi_k) - \frac{1}{2}\mu_k \Sigma_k^{-1}\mu_k' + \mathbf{x}\Sigma_k^{-1}\mu_k'.$$

QDA allows more flexibility on the variation of the classes and thus can be regarded as a generalization of LDA. Table 4.1 shows the accuracies of LDA and QDA when one only considers Petal Length and Petal Width. In the case of big data, QDA is expected to be more accurate if the covariance matrix can be properly estimated.

4.5 Sliced Inverse Regression (SIR)

In a fashion similar to LDA, one can also utilize the information based on the response while obtaining the maximal "variation." Instead of directly obtaining the maximal variation based on the input features (e.g., PCA), sliced inverse regression (SIR) obtains the linear combination of features resulting in the maximal variation of the expected value $E(X|y)$. (Related approaches from a machine learning standpoint involving the responses/targets are discussed in Chap. 8.)

Example 4.9 (SIR with Iris Data) SIR is commonly utilized when the response is a continuous variable. For a comparison to other method presented earlier, Fig. 4.6 shows the first two dimensions of the Iris data (by slicing the data into 5 subgroups for each species) run on SIR.   □

Criteria other than maximizing the variation of X consider the information in the response feature Y. Ker-Chau [8] proposed the method of *sliced inverse regression* for dimensionality reduction. The criterion is now to maximize the variation of the conditional expectation $E(X|y)$, instead of the variation of the covariates X. The directions retaining the maximal variation of $E(X|y)$ will better capture the change in response Y.

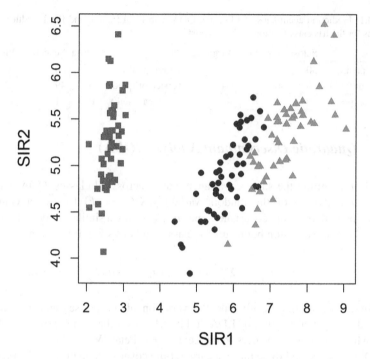

Fig. 4.6 Scatter plot on the first two dimensions of the Iris dataset via SIR

Example 4.10 (The Space Spanned by $X|Y$) In the simplest case, no error terms occur and the underlying model has the form

$$y = f(\boldsymbol{\beta}_1 \mathbf{x}'),$$

where β_1 is a pD row vector. The objective is to choose the right value of $\boldsymbol{\beta}_1$. One can rewrite the underlying model as

$$\boldsymbol{\beta}_1 \mathbf{x}' = f^{-1}(y)$$

assuming that f is invertible. Changing y, the $\boldsymbol{\beta}_1 \mathbf{x}'$ will change accordingly. Therefore, one only needs to examine the space spanned by $\mathbf{X}|Y$ to reveal the coefficient $\boldsymbol{\beta}_1$.   □

In a real world situation, the underlying model can be as complex as mentioned in Chap. 2, i.e.,

$$y = f(\boldsymbol{\beta}_1 \mathbf{x}', \boldsymbol{\beta}_2 \mathbf{x}', \dots, \boldsymbol{\beta}_K \mathbf{x}') + \varepsilon.$$

Then, following similar strategies, one only needs to study the space spanned by $E(X|Y)$ due to the error terms ε. The general procedure can be expressed as follows:

- *Slice data* into subdata sets based on the range of Y;
- Estimate the *inverse mean* $E(X|y)$ of each subdata;
- Obtain the direction which maximizes the variation of $E(X|y)$.

Practically, the number of slices that can be selected depends on the number of desirable reduced dimensions. For example, if only two dimensions are needed, the number of slices can be as small as 3. The reduction will converge quickly if more slices are utilized. For example, if the number of slices is 3 in the Iris dataset, the reduced dimensions via SIR will be similar to the reduced dimensions via LDA.

In summary, statistical DR exploits variations of the theme of PCA using statistical learning to fit parameters for classification and clustering problems, based only on predictor and/or target features.

References

1. Cichoski, A., Zdunek, R., Phan, A. H., & Amari, S. I. (2009). *Nonnegative matrix and tensor factorizations (applications to exploratory multi-way data analysis and blind source separation).* Wiley & Sons.
2. Ding, C., He, X., & Simon, H. D. (2001). On the equivalence of nonnegative matrix factorization and spectral clustering. In *Proc. SIAM Int'l Conf. Data Mining* (pp. 606–610).
3. Ding, Y., Lei, X., Liao, B., & Wu, F. X. (2022). Predicting miRNA-disease associations based on multi-view variational graph auto-encoder with matrix factorization. *IEEE Journal of Biomedical and Health Informatics, 26*(1), 446–457.
4. Fisher, R. A. (1936). The use of multiple measurements in taxonomic problems. *Annals of Eugenics, 7*(2), 179–188.
5. Gillis, N. (2020). *Nonnegative matrix and tensor factorizations.* SIAM-Society for Industrial and Applied Mathematics.
6. Hyvärinen, A., & Oja, E. (2000). Independent component analysis: Algorithms and applications. *Neural Networks, 13*(4–5), 411–430.
7. Lee, D. D., & Seung, H. S. (2001). Algorithms for non-negative matrix factorization. In *Advances in Neural Information Processing Systems 13: Proceedings of the 2000 Conference* (pp. 556–562).
8. Li, K. C. (1991). Sliced inverse regression for dimension reduction. *Journal of the American Statistical Association, 86*(414), 316–327.
9. Wang, C. Y., Gao, Y. L., Kong, X. Z., Liu, J. X., & Zheng, C. H. (2022). Unsupervised cluster analysis and gene marker extraction of scRNA-seq data based on non-negative matrix factorization. *IEEE Journal of Biomedical and Health Informatics, 26*(1), 458–467.

Chapter 5
Geometric Approaches

Nirman Kumar (iD)

Abstract It is often the case that data is encoded as numeric vectors and hence is naturally embedded in a Euclidean space, with a dimension equal to the number of features. After the classical PCA that fits a linear (flat) subspace so that the total sum of squared distances of the data from the subspace (errors) is minimized, any distance function in this space can be used to endow it with a geometric structure, where ordinary intuition can be particularly powerful tools to reduce dimensionality. The idea can be generalized by changing the flat space to obtain a possibly nonlinear curved object (a so-called manifold) that can be fitted to the data while trying to minimize the deformations of distances as much as possible. Four major methods of this kind are reviewed, namely MDS, ISOMAP, t-SNE, and random projections.

5.1 Introduction to Manifolds

Data is (or can be, even nonnumeric) often encoded as numeric vectors and hence can be naturally considered to "live" in a Euclidean space. The dimension of this space appears to be just the number of features in each data point. However, the data itself can be essentially lower-dimensional.

Example 5.1 Consider a synthetic 10D dataset generated by the following randomized process. We repeatedly take independent samples X_i, Y_i of two standard normal distributions $X_i, Y_i \sim N(0, 1)$ and set the jth datum in a data point $\mathbf{x}_i = (\mathbf{x}_{i_1}, \ldots, \mathbf{x}_{i_{10}})$ to be

$$\mathbf{x}_{ij} = 10jX_i + 2j^2Y_i + \varepsilon_{ij},$$

where $\varepsilon_{ij} \sim N(0, 10^{-4})$ is some small random error distributed normally as Gaussian noise with mean 0 and variance $\sigma^2 = 10^{-8}$.

N. Kumar (✉)
Computer Science, The University of Memphis, Memphis, TN, USA
e-mail: nkumar8@memphis.edu

© The Author(s), under exclusive license to Springer Nature Switzerland AG 2022
M. Garzon et al. (eds.), *Dimensionality Reduction in Data Science*,
https://doi.org/10.1007/978-3-031-05371-9_5

Arguably, the *real* dimensionality of the data is 2 rather than 10, but that is not apparent if one is just looking at the dataset without knowing exactly how it came about. The goal of dimensionality reduction is to discover this true lower dimension.

       □

Most humans have an intuitive sense of what *dimension* may mean, and these intuitive ideas may be good enough for informal discussions. However, Example 5.1 raises the question as to what is really meant by *dimension* of a dataset. Thus, a serious discussion of dimensionality reduction necessitates a deeper discussion and proper definition of this term. The traditional area of knowledge where it belongs is geometry, and mathematicians have developed conceptual frameworks to understand it, including some generalizations in subject areas such as linear algebra, algebraic geometry, differential geometry, and topology. The goal of this chapter is to examine these geometric concepts to understand *dimension* and use geometric ideas to develop effective dimensionality reduction techniques.

A geometric approach to dimensionality reduction strives to find fewer features that "preserve" essential information in the data for a given problem (e.g., shape), which eventually translates into various relationships among the distances across all data points, like in a sphere. In a nutshell, the geometric approach is to reduce the dimensionality while minimizing some appropriate function (e.g., a loss function) that quantifies discrepancies between a) distances between points in the given space and b) mapped points in the lower-dimensional space.

Example 5.2 PCA can be interpreted geometrically as fitting a linear (flat) subspace to the data (as shown in Fig. 3.1) so that the error defined as the total sum of squared distances of the data points from the subspace is minimized.   □

However, in general, the data may not be close to "flat." A curved object may fit the data better. In general, PCA will not be very successful in fitting a linear subspace of the "correct" dimension to the data.

Example 5.3 The 3D dataset of 200 random points on a spiral is depicted in Fig. 5.1. Any attempt to reduce dimensionality using PCA on this dataset will not yield any useful reduction since there is no lower-dimensional subspace that can fit the data points close enough.   □

This example raises the need to discuss two points. First, it should be clear that the notion of "dimension" requires a more careful definition, even if only theoretical. Intuitively, the dimension of the dataset in Example 5.3 is 3 since it does not lie in any plane. Moreover, even projecting to most planes may cause two different data points, far off from each other, to collapse onto the same point.

The second issue is how one should define the distance between data points in the general case for the purposes of dimension reduction, i.e., Euclidean distances may not always be the best choice. A geometric technique for DR by strictly preserving Euclidean distance between data points may not be successful, and alternative distance functions (defined in Sect. 1.2) may work better. In Example 5.3, it seems clear that this might be impossible to do without significantly distorting

Fig. 5.1 (a) A synthetic 3D dataset of 200 random points. (b) The data actually lies along a spiral curve (not part of the data) so data points can be parametrized by the distance from one end along the spiral and so it is really 1D

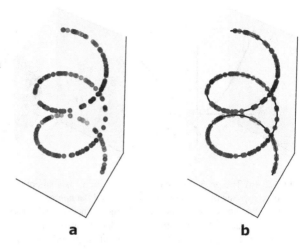

a   b

some distances. However, looking at it another way, the dimensionality of the data is not even 2. In fact, it is 1 if one agrees that distances between data points are the distances along the spiral curve. Under this interpretation, the data points on the spiral curve can be unrolled to a 1D line. The reason for this apparently tricky and difficult situation is that the data has a somewhat curved shape. Mathematicians have developed a conceptual apparatus to address curves, already present in elementary Calculus, with the concepts of tangent lines and derivatives to approximate curves with straight objects such as tangent lines, at least locally around a point. This property can be generalized to higher-dimensional situations where tangent planes or spaces must be used. Such smooth curved objects are technically referred to as *smooth manifolds*, and the areas of mathematics studying them are called *differential geometry* and its generalization, *topology* (to be defined further below).

Example 5.4 The ordinary *graph* of a differentiable function (one that has derivatives) such as $f(x) = x^2$ is an example of a smooth 1D manifold because for every point of the graph (t, t^2) one can fit a straight tangent line of slope $m = 2t$ that approximates the points and shape of the parabola at points sufficiently close to (t, t^2) regardless of the value of t, as illustrated in Fig. 5.2a. Likewise, as an idealization of the points on the surface of the Earth, a sphere is a smooth 2D manifold because at every point (x, y, z) one can find a flat 2D object, i.e., a *tangent plane* that does essentially the same for the sphere, as illustrated in Fig. 5.2b. □

The critical concept is thus that of a *neighborhood*. For the purpose of this book, the concept can simply be understood as a ball (without the boundary, as defined in Sect. 1.3).

Two subsets in two Euclidean spaces **E**, **E**′ are *smoothly homeomorphic* if and only if there exists a smooth bijective mapping

$$\phi : \mathbf{E} \longrightarrow \mathbf{E}'$$

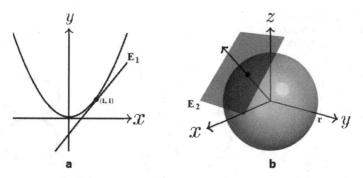

Fig. 5.2 (a) The *graph* of a parabola like $y = x^2$ is a smooth 1D manifold because the tangent line E_1 at any one of its points (x, x^2) is a good flat approximation within a small neighborhood of any point on the parabola. Likewise, (**b**) the surface of the unit sphere in 3D Euclidean space is a smooth 2D manifold because the tangent plane E_2 at any one of its points $(x, y, \pm\sqrt{r^2 - x^2 - y^2})$ is a good 2D flat approximation within a small neighborhood of the point on the sphere of radius r

with a smooth inverse between them that preserves neighborhoods, i.e., ϕ maps some neighborhood of every point $x \in E$ onto a neighborhood of the point $\phi(y) \in \mathbf{E}'$, for all $x, y \in \mathbf{E}$.

Example 5.5 Two circles (spheres) of radii 1 and 4 in the Euclidean plane (space) are smoothly homeomorphic by the mapping (e.g., $\phi(x, y) = (4x, 4y)$) in the first quadrant. However, a circle in the plane consisting of points at a set distance $|x, \mathbf{0}| = 1$ from the origin is not homeomorphic to an interval in the 1D line because such a ϕ cannot exist (being differentiable implies being continuous, and one cannot transform the circle into the interval without cutting it somewhere, i.e., introducing a discontinuity in the transformation). Likewise for a sphere in 3D Euclidean space and a 2D disk. □

A *d*D *smooth manifold* $\mathcal{M}$ is a subset of Euclidean space where every point x has a neighborhood that is smoothly homeomorphic to some neighborhood in the *d*D Euclidean space $\mathbf{R}^d$.

In other words, dD smooth manifolds are nice objects in Euclidean spaces in the sense that they have neighborhoods at every point that look like (i.e., are smoothly deformable without tearing or breaking) into flat neighborhoods of Euclidean spaces of some fixed Euclidean dimension d, despite the fact that they may be living in a higher-dimensional Euclidean space $\mathbf{R}^n$. The calculus of manifolds shows how to extend the existing mathematical techniques on flat spaces to curved spaces that may not be full Euclidean spaces, but only look locally like them.

Example 5.6 No one would argue that "locally," say within the distance one can walk in an hour, the Earth is nice and flat just like the surface of a table. Thus, the surface of the Earth can be thought of as a space that *locally* looks like a flat 2D plane. Of course, it is now known that this is an "illusion," but our ancestors found it very real and this only happens because the radius of the Earth is very large and

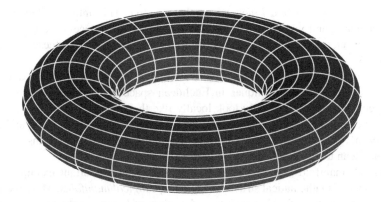

Fig. 5.3 A visualization of a torus, an idealization of the surface of a doughnut

one is only looking at a very small patch on its surface. What is then the "intrinsic" geometry of the Earth's surface? Should the surface of the Earth be considered 2D, which is so natural because it looks like a 2D plane *locally*, or 3D because it is defined as part of a 3D Euclidean space and intuitively one could not stretch out the whole surface of the Earth onto a 2-dimensional plane without "tearing" it? A manifold geometry would consider the surface of the Earth to be a 2D manifold, but not a plane. □

Example 5.7 An idealization of the surface of a doughnut (known in mathematics as a 2D *torus*) is also a smooth 2D manifold, as illustrated in Fig. 5.3. Although it is not smoothly homeomorphic (cannot be deformed) onto a sphere, it is surely smoothly homeomorphic to an ordinary coffee mug. The small flat patches provide local approximations to the curved surface. A small patch can be smoothly and bijectively mapped onto an open disk (i.e., an open 2D ball). □

Example 5.8 The surface of a cube in 3D Euclidean space has sharp corners, and as such, under the inherited metric from the ambient Euclidean space, it is not a smooth 2D manifold. (The transformation ϕ mapping a corner to a neighborhood in the 2D plane is not differentiable.) □

On a smooth manifold, one can use the notion of length of a shortest path between two points x, y and speak of the *geodesic distance* from x to y, which is called a *Riemannian metric*. Manifolds endowed with such a concept are called *Riemannian manifolds*. Thus, all smooth manifolds are also Riemannian manifolds. (Their technical definition refers to line integrals on the manifold along paths, but the conceptual idea is good enough here.) Indeed, one of the manifold learning algorithms (ISOMAP) described in Sect. 5.2 tries to approximate this process—it assumes that data lies on a Riemannian manifold and that around a small Euclidean neighborhood the geodesic distances are very well approximated by the Euclidean distances, so in a small neighborhood (defined using k-nearest neighbors of the points for some k), the Euclidean distances are used. Otherwise a shortest path algorithm is used on the k-nearest neighbor graph. This intuitive definition of Riemannian manifolds should hopefully demystify this process, at least on an intuitive level, to convey how DR works on "curved" datasets using these geometric ideas.

The crucial point is that the calculus of manifolds shows how to extend the existing mathematical techniques in Euclidean spaces to curved spaces that are non-Euclidean spaces, but only look locally like them. These matters are relevant to data science because even though data may appear high-dimensional and non-Euclidean, one may be able to analyze them with concepts and algorithms available for Euclidean techniques, even in lower dimensions.

Smooth manifolds such as the examples above are important examples of a larger class of mathematical structures called *topological manifolds*. More recently, mathematicians have found a way to make sense of homeomophisms without the notion of smoothness (differentiability) or measuring distances, by distilling the notion of neighborhoods to more general spaces, if one is willing to preserve certain more generic but essential constraints. The resulting structures are called *topological spaces* and *topological manifolds*, of which the (historically first) smooth manifolds discussed above are just important examples. The structures enable geometric concepts such as *continuity, connectedness*, and deformation to make sense in more general settings. Thus, every smooth manifold in the Euclidean space $\mathbf{R}^d$ is a topological space as is the surface of a sphere. Topology studies what properties of spaces remain unchanged under arbitrary homeomorphisms (i.e., continuous functions that are not necessarily smooth), such as the deformation of a sphere into a pointed cube in 3D space mentioned above. Homeomophisms are transformations that do not "tear" the space, although they may "bend" it. Properties such as the connectivity of the ball do not change, and as such it is a topological property. Topology is thus also known as *rubber sheet geometry*, and topologists are also described as mathematicians who cannot distinguish between a coffee cup and a donut! General topological spaces can be very complicated and weird as allowed by these abstract generalizations, but all of them are nice in the sense that they have neighborhoods at every point that are like flat Euclidean spaces of some fixed dimension d. More precisely, a topological d-manifold is a topological space such that every point has a neighborhood *homeomorphic* to $\mathbf{R}^d$, i.e., the neighborhood is continuously deformable to $\mathbf{R}^d$ in a one-to-one manner (bijectively). (There are other technical conditions that are parts of the definition that must be left out to keep the discussion manageable.)

Manifolds are important objects for DR. For example, points sampled from the curved surface of the 3D Chicago Bean shown in Fig. 5.4 can be locally (i.e., neighborhood-wise) conceptualized as flat, so the bean has intrinsic dimensionality 2 not 3. In doing a dimension reduction for this dataset, one would seek a mapping of the Bean's surface to the plane that is nice in any neighborhood of a point, i.e., it maps a patch on its surface to a disk in the plane. Thus, in this geometric view of DR, the dimension reduction is done "locally," but that is still consistent globally. The concept of manifold also helps clarify the notion of *geometric dimension*. First, let us consider *affine dimension*. Suppose for a given dataset, if all the data points

Fig. 5.4 The surface of the Chicago Bean is a 2D manifold

lie on a line (collinear), then the data is one-dimensional. If the data lies on a two-dimensional plane, then it is two-dimensional and so on (multicollinearity). The notion of dimension being used is the commonly used affine dimension, i.e., the smallest possible dimension of a subspace such that its translate contains the data points. PCA can be used to fit the smallest dimensional affine subspace possible to the data. However, this notion is restrictive since it assumes some "flatness" in the data. For example, if the data points lie on a circle situated in 3D space, it will be considered 2D, as the circle sits in a 2D plane; likewise, a torus cannot be 2D because it does not live in a 2D Euclidean plane. Yet, the notion of intrinsic dimension for a circle is 1D, more consistent with our intuition since every point can be described by a single number, namely its distance from some fixed point on the circle. Likewise for data on a torus is 2D (parametrizing along two orthogonal circles on the torus). This way of understanding the circle or the torus is referred to as a *parametrization*. The circle is also a 1D manifold, and the torus is a 2D manifold. Generalizing this example, if the data lies on a dD-manifold in a Euclidean space $\mathbf{R}^n$, the *dimensionality of the data is the dimension of the manifold, d, even if $d < n$*. Thus, intrinsic dimension can be thought of as the smallest number of independent parameters needed to "describe" the data, if only locally.

Further evidence of the importance of manifolds for data science comes from the following observation borne from experience, even for data that may be high-dimensional (i.e., with many features).

The Manifold Hypothesis
All points in every dataset usually lie on or close to a low-dimensional manifold.

The term "manifold" is being stretched here a bit (and so should not be taken too seriously) to mean precise mathematically defined manifolds (as in the examples above). Indeed, the data could lie, say on the union of two manifolds and still be considered low-dimensional. In addition, there is always the noise factor with real data. So, the statement is not an infallible fact (otherwise it would be a theorem), just an intuitive observation across datasets, a guiding principle. It should be rather interpreted to mean that the data "looks" like a mathematically defined low-dimensional manifold. Now, data is easy to understand and visualize when the manifold, such as the circle in the example above, is given in the way the data is encoded or organized (as part of the ontology, perhaps using a parametrization). However, data is usually just given as a table of numeric data points, not by a parametrization. *The goal of geometric methods for DR is to fit a manifold of the smallest possible dimension to the given data,* perhaps at the cost of losing some precision in passing to an approximation. How can such a manifold be constructed? Once that is done, how can it be parametrized? Such techniques are also known as manifold learning methods. They usually directly provide a mapping of the unknown manifold to a lower-dimensional Euclidean space, and this mapping also serves as a parametrization. The following subsections present a review of the most common methods. ([16] provide a more thorough treatment.)

5.2 Manifold Learning Methods

This section describes three major manifold learning methods, namely multi-dimensional scaling (MDS), isometric mapping (ISOMAP), and t-stochastic neighbor embedding (t-SNE). Two datasets will be used in running examples:

- The Swiss roll dataset is a synthetic set on a curved surface in 3D that models the surface of the common Swiss roll pastry. To generate this dataset, one first generates some random points (t, h) on the Euclidean 2D plane and then lifts them to 3D space by mapping them to $(t \cos(t), h, t \sin(t))$. Figure 5.5 shows such a dataset of size 1000.
- The *modified* National Institute of Standards and Technology (MNIST) dataset consists of 28×28 grayscale images of single handwritten ZIP codes (described in more detail in Sect. 11.5). A toy version of this dataset consisting of 1,797 data points (each a 8×8 64D binary image of a single digit from 0 to 9) was used to extract a random subsample of 597 data points, some of which are shown in Fig. 5.6 to illustrate the dimensionality reduction techniques described next.

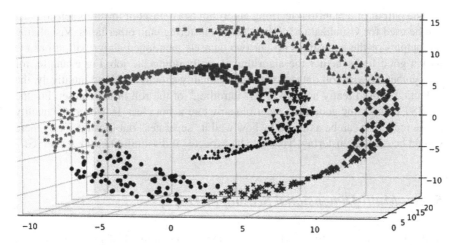

Fig. 5.5 The Swiss roll dataset with 1000 points. The shapes of the points indicate their position along the roll

Fig. 5.6 A subset of 120 images from the MNIST dataset

Dimensionality reduction here attempts to figure out a small number of variation parameters such as slope, height, and roundness of curves that contain enough information to identify the digit.

The basic philosophy of MDS and ISOMAP is to map the data points onto a lower-dimensional manifold such that the Euclidean distances between pairs of points reflect the "dissimilarity" between the original pairs of data points. The dissimilarities are explicitly given for MDS (assuming they are coming from the same problem domain), while ISOMAP assumes they lie on some higher-dimensional manifold and can be parametrized using the intrinsic geodesic distance on the manifold. The *t*-SNE algorithm computes distributions defined using the dissimilarity data and tries to match up these distributions in the reduction to the lower-dimensional space as closely as possible.

The output of a manifold learning algorithm is a dataset in lower dimension that can be used for visualization, classification, clustering, and other tasks. Visualizing the outputs of the algorithms on our two datasets mentioned above will afford the user a good feel whether the algorithm is really doing the job. For example, for the synthetic Swiss roll dataset, the dimensionality should be 2, and intuitively, the reduction should really amount to an "unrolling" of the roll in the dataset. For the MNIST dataset, the geometric manifold structure is not clear. However, the quality of the reduction can be assessed by how well it "separates" out the digits, i.e., there should be a good clustering. This is to be expected if dissimilarities are preserved.

5.2.1 Multi-Dimensional Scaling (MDS)

Multi-dimensional scaling (MDS) originated in the subfield of psychometrics in psychology. The term was coined by Torgerson, who also proposed the first method for solving the problem [23]. In psychometrics, a basic technique is to assign quantitative coordinates (scaling) to judgments of stimuli such as "object A is brighter than object B" or "A is twice as large as B." As observed by Torgerson, what seems to be a single-dimensional judgment, say "brighter" or "larger," may in fact be multi-dimensional, i.e., multiple factors may be involved in the qualitative judgment (hence, the M for *multi-dimensional* in MDS). An issue here is the reference or zero point in assessments such as "A is twice as large as B," as opposed to "A is larger than B." Torgerson calls these *comparative distances* and argues that after a shift by a constant c, such comparative distances can be interpreted as absolute distances. After this step is taken, the basic problem is whether, given some absolute pairwise distances between some objects (i.e., a distance matrix), one can assign coordinates so that these assessments are actually realized as distances between points in some Euclidean space. Such a realization would help in visualizing the set of objects. This is a very good example of MDS in action for the data scientist.

Example 5.9 Figure 5.7 shows a visualization of the Swiss roll dataset (obtained using the Python scikit-learn library, described in Sect. 11.5). From the visualization, it is clear that the method tried to flatten the Swiss roll to a 2D plane.   □

Example 5.10 Figure 5.8 shows a visualization of the MNIST digits dataset. Here, it seems that points corresponding to some digits such as "0" and "9" are well separated from the rest, but clearly similar to each other, while digits such as "8" and "5" are more spread out. Clearly, MDS does not produce a very good clustering here. (*t*-**SNE** produces a very clear visualization of the MNIST dataset below.)   □

Torgeson's problem of "scaling" is also known as the *metric MDS* problem and turns up in other fields such as marketing, sociology, biology, and data science. When the specified distance matrix actually arises from Euclidean distances, a procedure based on spectral methods can be used to find the mapping of objects to points in Euclidean space and is known as *classical MDS*. The algorithm will

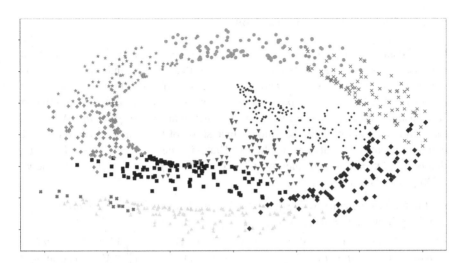

Fig. 5.7 The Swiss roll dataset mapped to 2D space by MDS

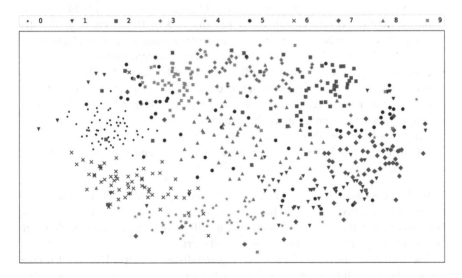

Fig. 5.8 The MNIST digits dataset sample mapped to a 2D space by MDS (obtained using the Python scikit-learn library)

succeed if distances are actually Euclidean, and then it is equivalent to the PCA algorithm. Surprisingly, the algorithm "works" (producing some output points) even if the given distances are not Euclidean, but clearly the distance between the produced points may not equal the desired input distances. Another approach based on an optimization view is also used here. These two approaches are discussed separately next.

5.2.1.1 Classical MDS: Spectral Approach

Given a dataset of size n to reduce into a set O of n mapped objects $\mathbf{x}_1, \ldots, \mathbf{x}_n$, their pairwise distances define a matrix $\mathbf{D} = [d_{ij}]$, so that $d_{ii} = 0$ and $d_{ij} = d_{ji}$ for all $1 \leq i, j \leq n$, where each $\mathbf{x}_i = (x_{i1}, \ldots, x_{ik})'$, is a point in pD space. For now, p remains undetermined. The unknowns $\mathbf{x}_1, \ldots, \mathbf{x}_n$ can be organized as a $k \times n$ matrix $\mathbf{X}$, as usual. The goal of MDS is to find $\mathbf{x}_i$s such that $\|\mathbf{x}_i - \mathbf{x}_j\| = d_{ij}$. As posed, the solution cannot possibly be unique since shifts of the points in a solution by a common vector $\mathbf{c}$ will also be a solution. Thus, the mean of the $\mathbf{x}_i$ should be zero, that is $\sum_{i=1}^{n} \mathbf{x}_i = \mathbf{0}$ or equivalently, $\sum_{i=1}^{n} x_{ji} = 0$ for $j = 1, \ldots, p$. To find the $\mathbf{x}_i$, MDS proceeds as follows:

- It computes the entry-wise squares of $\mathbf{D}$ (also known as the *Hadamard square*), i.e., the matrix $\mathbf{A} = (a_{ij})$, where $a_{ij} = d_{ij}^2$.
- It computes the matrix $\mathbf{B} = -\frac{1}{2}\mathbf{HAH}$, where $\mathbf{H} = \mathbf{I} - \frac{1}{n}\mathbf{11}'$ is the *centering matrix*, $\mathbf{I}$ is the $n \times n$ identity matrix, and $\mathbf{1} = [1, 1, \ldots, 1]' \in \mathbf{R}^n$ is a column of 1's. (Altogether these steps transform the distance matrix to the *Gram matrix* of the unknown points, i.e., if the desired condition $\|\mathbf{x}_i - \mathbf{x}_j\| = d_{ij}$ holds true and the mean of the $\mathbf{x}_i$ is 0, then it will be true that $\mathbf{X}'\mathbf{X} = \mathbf{B}$, i.e., the (i, j)-th entry is not the distance between $\mathbf{x}_i, \mathbf{x}_j$, but their dot product $\mathbf{x}_i'\mathbf{x}_j$. The advantage of this is that $\mathbf{X}$ can be further recovered by a spectral decomposition.)
- It finds the spectral decomposition $\mathbf{B} = \mathbf{V}\Lambda\mathbf{V}'$ of $\mathbf{B}$, where $\Lambda = \mathbf{D}(\lambda_1, \ldots, \lambda_n)$ is the diagonal matrix of eigenvalues of $\mathbf{B}$ and $\mathbf{V} = [\mathbf{v}_1, \ldots, \mathbf{v}_n]$ is the matrix of eigenvectors normalized such that $\mathbf{v}'\mathbf{v} = 1$ (as described in Sect. 11.2), and the eigenvalues have been arranged in decreasing order $\lambda_1 \geq \lambda_2 \geq \ldots \geq \lambda_n \geq 0$. Let $p \leq n$ be the number of nonzero eigenvalues.
- It finds $\mathbf{B} = \mathbf{V}_1\Lambda_1\mathbf{V}_1'$, where $\Lambda_1 = \mathsf{diag}(\lambda_1, \ldots, \lambda_p)$ and $\mathbf{V}_1 = [\mathbf{v}_1, \ldots, \mathbf{v}_p]$.
- It returns $\mathbf{X}$ as $\mathbf{X} = \mathbf{V}_1\Lambda_1^{1/2}$, where $\Lambda_1^{1/2} = \mathsf{diag}(\lambda_1^{1/2}, \ldots, \lambda_p^{1/2})$.

In this procedure, the sign of the eigenvectors $\mathbf{v}_i$ could be all flipped (corresponding to a reflection across the origin). Thus while centering removes some indeterminacy in the solution, reflections can still leave a choice of sign. However, this is not really an issue, as either choice will work.

The algorithm above can be used on a given distance matrix $\mathbf{D}$, even if it does not arise from Euclidean distances. What will happen in this case is that the matrix in step 2 above will not necessarily be a Gram matrix. However, if it is positive semi-definite, the algorithm would still "go through" producing the desired matrix $\mathbf{X}$. On the other hand, if $\mathbf{B}$ is not positive semi-definite (this is equivalent to some eigenvalue being negative or complex), one approach [18] is to make $\mathbf{B}$ positive semi-definite by starting with a new distance matrix $\mathbf{D}$, after adding the same constant c to each entry d_{ij} with $i \neq j$. This additive constant can always be found, as shown by Messick and Abelson [18]. Another approach is to neglect the negative eigenvalues (and corresponding eigenvectors) [5]. In this case, MDS produces some sort of approximate solution to the problem, i.e., it will not be the case that $\|\mathbf{x}_i - \mathbf{x}_j\| = d_{ij}$, but hopefully so that $\|\mathbf{x}_i - \mathbf{x}_j\| \approx d_{ij}$. Another remark is that the dimension of the space for the points x_i does not necessarily have to be the number of nonzero eigenvalues. Any number $k \leq p$ less than the number of nonzero eigenvalues will do. (This is analogous to choosing the top k principal components in PCA.)

5.2.1.2 Metric MDS: Optimization-Based Approach

In this case, the given distance matrix $\mathbf{D}$ is now called the matrix of dissimilarities. They do not have to be Euclidean distances between the n points. In fact, they may be non-Euclidean metric distances, dissimilarities, or even not arising from a metric. Yet, one wants to find points $x_1, \ldots, x_n$ such that now $\|x_i - x_j\| \approx d_{ij}$, the best one can hope for under these conditions. The matrix $\mathbf{X} = [x_1, \ldots, x_n]$ is termed a *configuration*. Of course, if d_{ij} *actually arises from Euclidean distances*, then one might get equality for some configuration. What about the dimension of the space k in which the points are supposed to lie? Usually, k is also given as input or chosen to be something small, e.g., 2 or 3 since MDS is usually used for visualization of the data. The problem can thus be formulated as an optimization problem for a quantity $\sigma(\mathbf{X})$ called the *stress* of (unknown) configuration $\mathbf{X}$ and given by

$$\sigma(\mathbf{X}) = \sum_{i<j} w_{ij}(d_{ij} - \|x_i - x_j\|)^2,$$

where w_{ij} are the entries of a known matrix of weights. The weights are usually chosen simply as $w_{ij} = 1$ for all $i < j$, but more generally, they can be chosen so as to give higher weights for some pairs i, j, and lesser for others. In particular, in many data analysis situations, some of the d_{ij} may be completely unknown or not very trustworthy. In that case, one can set $w_{ij} = 0$. Thus, it is clear that an "ideal solution" to the problem would achieve $\sigma(\mathbf{X}) = 0$. Therefore, it is natural to define the solution $\mathbf{X}$ as

$$\mathbf{X} = \arg \min_{\mathbf{X} \in \mathbf{R}^{k \times n}} \sigma(\mathbf{X}).$$

In other words, MDS chooses an $\mathbf{X}$ that minimizes the stress. The stress $\sigma(\mathbf{X})$ seems like a complicated function to minimize. In fact, different definitions of stress may be used. It is interesting to observe that one such choice would, in particular, lead to the same minimization problem as PCA if the given objects are themselves points in $\mathbf{R}^d$, the dissimilarities are the distances, and $\mathbf{X}$ is required to be a projection of the data points onto some subspace.

There are also many ways to minimize the stress. SMACOF is a strategy to minimize a supposedly "complicated" function $f : \mathbf{R}^d \to \mathbf{R}$. The strategy works as follows. Given a guess $x^{(t)}$ for the argument minimizing f (perhaps chosen randomly to begin with), SMACOF finds a *majorizing* function $g : \mathbf{R}^d \times \mathbf{R}^d \to \mathbf{R}$ that is somewhat of a "simpler function" to minimize, i.e., such that $f(x) \leq$

$g(x, x^{(t)})$, $f(x^{(t)}) = g(x^{(t)}, x^{(t)})$ for all x in the domain and that can be minimized to produce a *better estimate* $x^{(t+1)}$ for the minimizer of f. The following chain of inequalities,

$$f(x^{(t+1)}) \leq g(x^{(t+1)}, x^{(t)}) \leq g(x^{(t)}, x^{(t)}) = f(x^{(t)}),$$

referred to as the *sandwich inequality*, shows that the minimum value of $g(x, x^{(t)})$ is sandwiched between $f(x^{(t+1)})$ and $f(x^{(t)})$. The algorithm replaces $x^{(t)}$ by $x^{(t+1)}$ and iterates the update until $f(x^{(t)}) - f(x^{(t-1)}) < \varepsilon$ for some fixed (small) constant ε, or until a certain number of iterations is reached. Although SMACOF can lead to local minima (as shown by Groenen and Heiser [9]), local minima are more likely to occur in lower-dimensional solutions (for very small values of k).

How can the SMACOF strategy help MDS? Well, it can be shown that the stress $\sigma(\mathbf{X})$ to be minimized can be re-written as

$$\sigma(\mathbf{X}) = \sum_{i<j} w_{ij}(d_{ij} - \|\mathbf{x}_i - \mathbf{x}_j\|)^2 = \eta_D^2 + \text{Tr}(\mathbf{X}'\mathbf{V}\mathbf{X}) - 2\text{Tr}(\mathbf{X}'\mathbf{F}(\mathbf{X})\mathbf{X}),$$

where $\text{Tr}(\cdot)$ is the trace operator (for square matrices), the value

$$\eta_D^2 = \sum_{i<j} w_{ij} d_{ij}^2$$

is a constant, the matrix $\mathbf{V}$ is an $n \times n$ matrix given as

$$[\mathbf{V}]_{ii} = \sum_{j \neq i} w_{ij}, \quad [\mathbf{V}]_{ij} = -w_{ij} \text{ for } i \neq j,$$

and $\mathbf{F}$ is a matrix function mapping matrix $\mathbf{X} = [\mathbf{x}_1, \ldots, \mathbf{x}_n]$ to another matrix $\mathbf{F}(\mathbf{X})$ defined entry-wise as

$$[\mathbf{F}(\mathbf{X})]_{ij} = \begin{cases} \frac{w_{ij} d_{ij}}{\|\mathbf{x}_i - \mathbf{x}_j\|} & \text{if } \|\mathbf{x}_i - \mathbf{x}_j\| \neq 0 \\ 0 & \text{if } \|\mathbf{x}_i - \mathbf{x}_j\| = 0. \end{cases}$$

Letting $\rho(\mathbf{X}) = \text{Tr}(\mathbf{X}'\mathbf{F}(\mathbf{X})\mathbf{X})$ and

$$\eta^2(\mathbf{X}) = \sum_{i<j} w_{ij}^2 \|\mathbf{x}_i - \mathbf{x}_j\|^2 = \text{Tr}(\mathbf{X}'\mathbf{V}\mathbf{X}),$$

it turns out that

$$\sigma(\mathbf{X}) = \eta_D^2 + \eta^2(\mathbf{X}) - 2\rho(\mathbf{X})$$

is actually a convex quadratic function since it is a sum of weighted squares of distances. The complicated part is $\rho(\mathbf{X})$, and the majorization strategy tries to separate it off. It can be shown that the following function is a good choice to majorize the stress,

$$T(\mathbf{X}, \mathbf{Y}) = \eta_D^2 + \eta^2(\mathbf{X}) - 2\tilde{\rho}(\mathbf{X}, \mathbf{Y}),$$

where $\tilde{\rho}(\mathbf{X}, \mathbf{Y}) = \text{Tr}(\mathbf{X}'\mathbf{F}(\mathbf{Y})\mathbf{Y})$. Then, the SMACOF strategy will minimize the stress starting from any configuration $\mathbf{X}^{(0)}$ (at $t = 0$) by iterating the following steps:

- Sets $Y = \mathbf{X}^{(t)}$ and computes the stress $\sigma(\mathbf{X}^{(t)})$
- Stops if $t > 0$ and $|\sigma(\mathbf{X}^{(t)}) - \sigma(\mathbf{X}^{(t-1)})| < \varepsilon$; else computes $\mathbf{X}^{(t+1)}$ by minimizing $T(\mathbf{X}, \mathbf{Y})$ over all $\mathbf{X}$

It turns out that some linear algebra shows that the minimum can be expressed as

$$\mathbf{X}^{(t+1)} = \mathbf{V}^+\mathbf{F}(\mathbf{Y})\mathbf{Y},$$

where $\mathbf{V}^+$ is the Moore–Penrose inverse,

$$\mathbf{V}^+ = \left(\mathbf{V} + \frac{1}{n}\mathbf{1}\mathbf{1}'\right)^{-1} - \frac{1}{n}\mathbf{1}\mathbf{1}'.$$

The transformation, $\mathbf{X}^{(t+1)} = \mathbf{V}^+\mathbf{F}(\mathbf{Y})\mathbf{Y}$, is known as the *Guttman transform* of configuration $\mathbf{Y}$ [10]. For the simple case often used in practice, where $w_{ij} = 1$ for all $i \neq j$, the Guttman transform simply becomes $\mathbf{X}^{(t+1)} = \frac{1}{n}\mathbf{F}(\mathbf{Y})\mathbf{Y}$.

In further work, Shepard [20] extended MDS to the *nonmetric MDS* case where only rankings of the distances are given, instead of precise distance information between pairs of objects. So, for each object A, one can rank the other objects in some ordering. Then, using a monotonic transformation function f, one can map such orderings to absolute distances and fall back to the metric case of MDS. ([5] give more details.)

The MDS procedure is an important method because it solves a very general problem. Also, it can be used in other manifold learning methods, such as ISOMAP.

5.2.2 Isometric Mapping (ISOMAP)

An *isometric mapping* (ISOMAP) is a mapping that preserves distances. It was introduced by Tenenbaum et al. [22]. If one could fit a low k-dimensional manifold to the data or come very close to it (a reasonable assumption in sight of the Manifold Hypothesis), the ISOMAP algorithm's basic philosophy is to assume that "learning" this manifold could help data analysis and that the distances between the points are the "intrinsic" geodesic distances between the points on the manifold. While this seems reasonable and intuitive, there are at least three problematic issues that must be considered:

- What is given is a sample of points and not the approximating manifold "equations" (for example, algebraic equations that define a surface, like (x, x^2) in 1D). How are such geodesic distances to be computed?
- Even if such distances could be computed, can the dimension be reduced while preserving the distances?
- Even if these two issues are addressed, how is an actual DR mapping to a lower-dimensional space to be done?

Intuitive justifications to address these issues are given below, but first an example is due to see ISOMAP in action.

Example 5.11 Figure 5.9 shows ISOMAP's output when applied to the MNIST dataset described in Sect. 11.5. As can be seen, ISOMAP does an excellent job. It essentially unrolls the Swiss roll flat onto a plane, while preserving the distances nicely, just like unrolling a piece of paper with the data rolled up into a cylinder (the manifold) back to a flat sheet of paper (the 2D Euclidean space).

Example 5.12 Figure 5.10 shows ISOMAP's output when applied to the MNIST dataset described in Sect. 11.5. Although a manifold structure is not obvious in the data, ISOMAP is comparable to MDS in performance. Some of the digits are nicely clustered, although others are spread out.

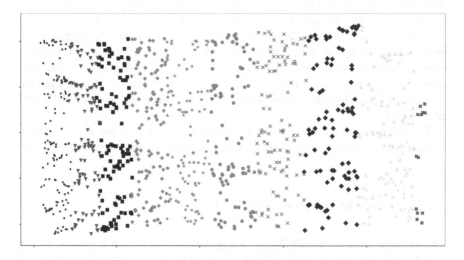

Fig. 5.9 The Swiss roll dataset mapped to 2D space by ISOMAP

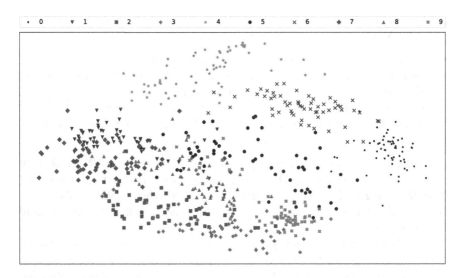

Fig. 5.10 The MNIST digits dataset sample mapped to 2D space by ISOMAP

To address the second issue first, at least at a somewhat intuitive level, i.e., without working through any mathematical definitions, a deep theorem of Nash [19] shows that:

Any Riemannian manifold of dimension k (initially lying in some higher- dimensional space $\mathbf{R}^d$, possibly with $d \gg k$) can be embedded isometrically into Euclidean space of dimension $2k + 1$.

Thus, at least in principle, one can rest assured that such a dimensionality reduction is possible. Now, the first issue mentioned above looks like a difficult question. In particular, even if the manifold is "known" analytically (e.g., by as some algebraic equations defining a surface), computing geodesic distances requires integrals and higher-dimensional Calculus. Being given just a sample of points, how can such distances be computed? Well, taking a practical approach and recalling that a manifold looks "locally flat," then the distances between nearby points can be simply computed using the distance formula for the Euclidean space. For farther-off points, the usual definition of a geodesic—that of a shortest path on the manifold between the points, can be used. In the absence of the manifold itself, such shortest path will be defined on an appropriate neighborhood weighted graph with the given data points as vertices. The neighbors of a vertex are vertices lying "close enough" so that the distance between them can be defined as the Euclidean lengths of the edges (as weights). A shortest path algorithm is then used to compute the distance between two arbitrary vertices of the graph. The neighborhood graphs are usually defined by selecting a small integer $\varepsilon > 0$ and ε nearest neighbors of a data point as the neighborhood of the point. (Alternatively, one could use all neighbors within a distance ε of a given data point.)

One obtains the so-called ε-neighborhood graph G_ε. With this parameter in place, ISOMAP proceeds as follows. It:

- Weighs each edge (u, v) with its distance $\|u - v\|$ in $\mathbf{R}^d$.
- Computes shortest path distances between every pair of vertices using any all-pair shortest path algorithm, e.g., Floyd–Warshall's.
- Returns the resulting $n \times n$ distance matrix D.
- Returns an appropriate value of the reduced dimension t using MDS on the distance matrix D.

ISOMAP has a worst-case run-time of $O(n^3)$. Using the ε-NN graph with $\varepsilon neighbors$, one can also use Dijkstra's algorithm with Fibonacci heaps to reduce it to $O(\varepsilon n^2 \log n)$ time. There are even faster approaches exploiting the sparse structure of the neighborhood graph (as in [15]). The algorithm answers the third issue raised above.

It has been shown theoretically that if the data indeed lies on a low-dimensional manifold whose geometry is that of a convex region of space and if the data points are sufficiently dense on this manifold, then ISOMAP returns a distance matrix that is close to the actual geodesic distances and, in the limiting case, converges to it. How dense the points should be for this to happen depends on some geometric parameters of the manifold. Even though the parameters are not known to the algorithm, this is an important theoretical result since it assures us that the method is "faithful" to the geodesic distances, i.e., sufficiently dense data would lead to the correct distance computation. The actual result is in fact quantitative. (Further details can be pursued in [22].)

The ISOMAP algorithm is very good at discovering nonlinear structures in the data. Since ISOMAP uses MDS in its last step, the stress loss function used in MDS can also serve to estimate the quality of the dimensionality reduction by ISOMAP. An alternative measure that can be used for PCA MDS and ISOMAP is the residual variance,

$$1 - R(D, D_X)^2,$$

where D is the distance matrix, D_X is the matrix of actual distances once the dimensionality reduction is done, and R is the standard linear correlation coefficient computed over the entries of D and D_X.

One problem with MDS, and hence ISOMAP, is that they are extremely slow in practice because the optimization is expensive, and so are the distance computations. These issues are somewhat addressed by a variant of ISOMAP called Landmark ISOMAP [21].

5.2.3   t-Stochastic Neighbor Embedding (t-SNE)

The *t*-**SNE** method by van der Maaten and Hinton [17] is a complex method that builds over and addresses the shortcomings of its precursor, the *stochastic neighbor embedding*(SNE) method by Hinton and Roweis [11]. So SNE is described first, pointing out the problems identified and then showing how *t*-**SNE** addresses them. In the process, the meaning of the somewhat intriguing *t* in *t*-**SNE** will be revealed. Overall, *t*-**SNE** is considered the state-of-the-art in geometric DR and visualization algorithms.

Example 5.13 Figure 5.11 shows a visualization of the Swiss roll dataset by *t*-**SNE**. *t*-SNE does an excellent job of clustering the like points together and ensuring good inter-cluster separation.   □

Example 5.14 Figure 5.12 shows a visualization of the MNIST dataset by *t*-**SNE**. In this case, it separates out the clusters corresponding to the points for each digit quite nicely. It seems to be doing a much better job, compared to MDS and ISOMAP.   □

Generally, given n data points $x_1, x_2, \ldots, x_n$ in some higher-dimensional space $\mathbf{R}^d$, the SNE algorithm associates to each point x_i (a base point) a probability distribution P_i obtained from the points in the dataset. These distributions are in general different for each of the points since they depend on the distances from the base point to the other points. More specifically, P_i is a Gaussian distribution

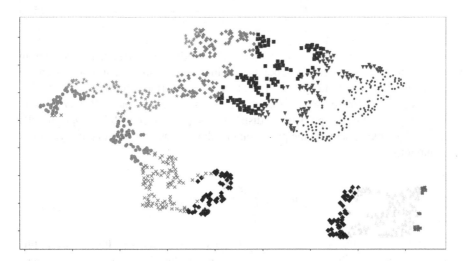

Fig. 5.11   The Swiss roll dataset mapped to 2D space by *t*-**SNE**

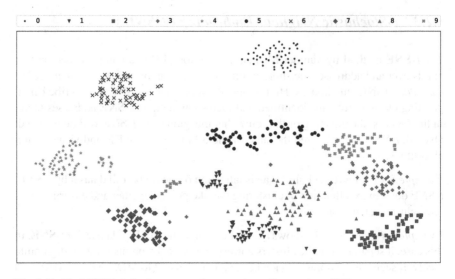

Fig. 5.12 The MNIST digits dataset sample mapped to ndS space by t-SNE

centered at base point $\mathbf{x}_i$ and determined by some standard deviation σ_i to be determined later. A weight for another point $\mathbf{x}_j$ ($j \neq i$) is set as

$$P_{j|i} = \frac{e^{-\|\mathbf{x}_i - \mathbf{x}_j\|^2 / 2\sigma_i^2}}{\sum_{k \neq i} e^{-\|\mathbf{x}_i - \mathbf{x}_k\|^2 / 2\sigma_i^2}}, \tag{5.1}$$

while the value $P_{i|i}$ is set to 0 since the method is interested in modeling dissimilarities between different points. It is important to notice that P_i assigns more probability measure to the points closer to $\mathbf{x}_i$ than to farther-off points.

Now, let $\mathbf{y}_i$ denote the point in the lower-dimensional space $\mathbf{R}^k$ that $\mathbf{x}_i$ is to be mapped to. Similar probability distributions Q_i with probabilities $q_{j|i}$ for the mapped points are set with the provision that the variance for the ith point $\mathbf{y}_i$ equals $\frac{1}{2}$ and thus

$$q_{j|i} = \frac{e^{-\|\mathbf{y}_i - \mathbf{y}_j\|^2}}{\sum_{k \neq i} e^{-\|\mathbf{y}_i - \mathbf{y}_k\|^2}}.$$

The variance in the lower-dimensional space is constant ($\frac{1}{2}$) for all the points but may possibly differ from the variances σ_i^2 in the higher- dimensional space. Therefore, even if SNE maps points from $\mathbf{R}^d$ to $\mathbf{R}^d$, it will not necessarily map them to the same points.

The SNE algorithm attempts to find the points $\mathbf{y}_i$ so that the sum of the Kullback–Leibler (KL) divergence (a measure of the dissimilarity of two probability distributions) of the distributions P_i and Q_i is minimized, i.e., so that

$$C = \sum_i KL(P_i \| Q_i) = \sum_i \sum_j p_{j|i} \log \frac{p_{j|i}}{q_{j|i}}$$

is as small as possible. However, this KL divergence is not a distance function between distributions because it is asymmetric and does not satisfy the triangle inequality. Lack of symmetry causes this loss function to penalize close together points by mapping them to far-off points more than vice versa. Ideally, one would like to have a perfect match: $q_{j|i} = p_{j|i}$, but this does not usually happen; hence, the "penalty" term $p_{j|i} \log \frac{p_{j|i}}{q_{j|i}}$ in C. The loss function is affected much more when $p_{j|i}$ is small and $q_{j|i}$ is large than vice versa. Intuitively, this means that the SNE cost function is really trying to preserve the *local structure* of the data, ignoring the *global structure* to some extent.

Now, how to choose σ_i? A single value of σ_i is probably not optimal for all data points. Usually, smaller values of σ_i are more appropriate in denser regions than in sparser regions of the data. Using the Shannon entropy $H(P_i)$ (defined in Sect. 6.1) of the distribution P_i, the quantity

$$Perp(P_i) = 2^{H(P_i)},$$

is called as the *perplexity* of the distribution P_i. It increases monotonically with σ_i. The perplexity can be interpreted as *a smooth measure of the effective number of neighbors*; typical values range between 5 and 50 [17]. In order to set σ_i, SNE requires the user to specify a perplexity value and then performs a binary search for σ_i such that the $Perp(P_i)$ matches the given value.

The minimization of the loss function C is done by gradient descent since the gradient is

$$\frac{\partial C}{\partial \mathbf{y}_i} = 2 \sum_j (p_{j|i} - q_{j|i} + p_{i|j} - q_{i|j})(\mathbf{y}_i - \mathbf{y}_j).$$

This gradient has a nice physical interpretation, as the resultant force exerted by a set of springs between $\mathbf{y}_i$ and other points $\mathbf{y}_j$. The spring between $\mathbf{y}_i$ and $\mathbf{y}_j$ exerts a force on $\mathbf{y}_j$ along the direction $\mathbf{y}_i - \mathbf{y}_j$, and it causes an attraction or repulsion based on the mismatch value $(p_{j|i} - q_{j|i} + p_{i|j} - q_{i|j})$. For example, if $\mathbf{x}_i, \mathbf{x}_j$ are very close, $p_{j|i}$ and $p_{i|j}$ are large. If $\mathbf{y}_i, \mathbf{y}_j$ are far apart, i.e., $q_{j|i}$ and $q_{i|j}$ are small, then the force,

$$(p_{j|i} - q_{j|i} + p_{i|j} - q_{i|j})(\mathbf{y}_i - \mathbf{y}_j),$$

is attractive and tries to bring $\mathbf{y}_j$ closer to $\mathbf{y}_i$. To initialize the gradient descent algorithm, n points are sampled randomly in $\mathbf{R}^k$ from an isotropic Gaussian distribution with mean at the origin, i.e., a multivariate distribution with marginals independent and Gaussian with the same standard deviation and mean 0. Then the $\mathbf{y}_i$s are repeatedly updated as

$$\mathbf{y}_i^{(t)} \leftarrow \mathbf{y}_i^{(t-1)} + \eta \left(\frac{\partial C}{\partial \mathbf{y}_i} \right)_{\mathbf{y}_i = \mathbf{y}_i^{(t-1)}} + \alpha(t)(\mathbf{y}_i^{(t-1)} - \mathbf{y}_i^{(t-2)}),$$

where η is the learning rate and $\alpha(t)$ is the momentum at iteration t. The effect of the momentum term is to effectively add a decaying sum of the previous gradients. Also, in early iterations, Gaussian noise is added after each iteration. Gradually reducing the variance of this noise (akin to *simulated annealing*) helps the algorithm drive the candidate solution out of "wells" about local minima. The optimization is thus quite complex, and in practice, one runs it several times to find results for DR and data visualization.

The variant of SNE, the t-distributed SNE algorithm (t-**SNE**), differs from SNE as follows. First, it breaks asymmetry between $p_{i|j}$ and $p_{j|i}$ by introducing $p_{ij} = (p_{i|j} + p_{j|i})/2n$; second, the q_{ij} are defined using Student's t-distribution as

$$q_{ij} = \frac{(1 + \|\mathbf{y}_i - \mathbf{y}_j\|^2)^{-1}}{\sum_{k \neq l}(1 + \|\mathbf{y}_k - \mathbf{y}_l\|^2)^{-1}}, \tag{5.2}$$

while keeping p_{ii} and q_{ii} equal to 0. Finally, instead of using different distributions P_i, Q_i one for each input and base point, it uses a single common distribution called P on $[n] \times [n]$ and Q (for the mapped points).

The cost function is now the KL divergence given by

$$C = KL(P\|Q) = \sum_i \sum_j p_{ij} \log \frac{p_{ij}}{q_{ij}}.$$

The reasons for the different choices made by the t-SNE algorithm are:

1. the $P_{i|j}$ and $P_{j|i}$ to P_{ij} are now symmetrized. Alternatively, this can be done, for example, by setting,

$$P_{ij} = \frac{e^{-\|\mathbf{x}_i - \mathbf{x}_j\|^2 / 2\sigma^2}}{\sum_{k \neq l} e^{-\|\mathbf{x}_k - \mathbf{x}_l\|^2 / 2\sigma^2}}.$$

However, if one of the points $\mathbf{x}_i$ is an outlier too far from the other points, the probabilities P_{ij} are all very small, and therefore the location of $\mathbf{y}_i$ does not affect the loss function much. The proposed symmetrization $P_{ij} = (P_{j|i} + P_{i|j})/2n$ ensures $\sum_j P_{ij} > 1/2n$ for all points $\mathbf{x}_i$, so that each point has some reasonable effect on the cost function.

2. This type of dimensionality reduction runs into the so-called *crowding problem*, i.e., in high dimension d, the volume of a ball of radius r scales as r^d. This implies that if the points are packed inside such a ball, say all at least $r/2$ from each other, one could pack about 4^d such points inside the ball, a nice and tight cluster (a "crowd") in the input data near x_i. In order to map this to a lower dimension k, one cannot pack all of the crowd inside a ball of radius r because such a ball will only contain about 4^k such points. Thus, *any* mapping to $\mathbf{R}^k$ will push some points farther off from the center of the ball, i.e., many distances will be much larger than, say r. Going back to the spring metaphor, there are many springs now that are very stretched, and therefore exert a large force on y_i. To be in equilibrium, y_i must be pushed somewhat to the center. Thus, the centers of all such tight clusters get pushed to the center and therefore closer to each other. This is counterproductive since now different clusters start to merge into each other. To alleviate this challenging problem, the *t*-**SNE** algorithm employs a heavy tailed distribution, like the Student's t with 1 degree of freedom (i.e., the Cauchy distribution). In a higher-dimensional space, a Gaussian is used to define the probabilities. By using a heavy tailed distribution in the lower-dimensional space, a moderate distance is allowed within the cluster to be faithfully modeled by a large distance in the lower-dimensional space. Thus, somewhat the tight cluster balloons *naturally* into a larger cluster in low dimensions. The net effect is that the attractive forces are not too large and do not force y_i too much to the center. The resulting map clusters the data much better.

The gradient of the loss function is now

$$\frac{\partial C}{\partial \mathbf{y}_i} = 4 \sum_j (p_{ij} - q_{ij})(\mathbf{y}_i - \mathbf{y}_j)(1 + \|\mathbf{y}_i - \mathbf{y}_j\|^2)^{-1}. \tag{5.3}$$

There are other reasons too for using the Student t-distribution. It has been shown that it is an infinite mixture of Gaussians [12] and is therefore closely related to the Gaussian distribution. It is also faster to evaluate since no exponentials are involved.

The final form of t-**SNE** requires the perplexity $Perp$, the maximum number of iterations T, a learning rate η, and a momentum $\alpha(t)$. It then proceeds as follows:

- It computes the $p_{j|i}$ using Eq. 5.1 and perplexity $Perp$ and sets $p_{ij} = (p_{j|i} + p_{i|j})/2n$.
- It samples an initial solution $\mathbf{Y}^{(0)} = (\mathbf{y}_1^{(0)}, \ldots, \mathbf{y}_n^{(0)})$, where each $\mathbf{y}_i^{(0)}$ is sampled from the standard normal distribution $N(0, 10^{-4}I)$.
- It repeats:

  - It computes the q_{ij} using Eq. 5.2.
  - It computes the loss function C and its gradient using Eq. 5.3.

– It updates

$$\mathbf{Y}^{(t)} = \mathbf{Y}^{(t-1)} + \eta \left(\frac{\partial C}{\partial \mathbf{Y}} \right)_{\mathbf{Y}=\mathbf{Y}^{(t-1)}} + \alpha(t)(\mathbf{Y}^{(t-1)} - \mathbf{Y}^{(t-2)}),$$

where $\frac{\partial C}{\partial \mathbf{Y}} = (\frac{\partial C}{\partial \mathbf{y_1}}, \ldots, \frac{\partial C}{\partial \mathbf{y_n}})$,

for $t = 1, 2, \ldots, T$.

Some tricks can be used to improve it, but they become too technical for the goal of this chapter. [17] has more details.

5.3   Exploiting Randomness (RND)

The geometric methods above were generally to map points to a lower-dimensional space such that the pairwise distances (dissimilarities) are preserved. Surprisingly for Euclidean distances, for any given finite set of points, there always exists a map that is simply a *projection* to a lower-dimensional subspace that approximately preserves *all* the interpoint distances! Moreover, the target dimension, i.e., the dimension of the space to which the projection is done, is independent of the space in which the given points lie. It only depends on n, the number of points, and, of course, how accurately the distances need to be preserved.

Example 5.15 In the *word document model*, a document is usually represented as a bag of words. Each document is a dD vector where d is the number of dictionary words (or a subset of such words of interest) and the ith entry is the count of such words in the document. A collection of n documents can be arranged as a dataset in a $d \times n$ matrix $\mathbf{D}$, where each column vector represents a document. The similarity of two documents represented by vectors $\mathbf{x}, \mathbf{y}$ is their dot product (defined in Sect. 11.2) denoted $\mathbf{x}' \cdot \mathbf{y}$ (or just $\mathbf{x}'\mathbf{y}$). Computing all such similarities would take about $O(n^2 d)$ time, which can be prohibitively large if d is very large.

To speed up this method, one way is to project all vectors $\mathbf{x}$ to $\bar{\mathbf{x}}$ in a much lower kD subspace ($k \ll d$) such that the Euclidean distances are approximately preserved, i.e., $\|\bar{\mathbf{x}} - \bar{\mathbf{y}}\| \approx \|\mathbf{x} - \mathbf{y}\|$. Therefore, so would be the dot products to within some additive error, i.e., $\bar{\mathbf{x}}'\bar{\mathbf{y}} \approx \mathbf{x}'\mathbf{y}$.

This section shows that a random projection to a lower-dimensional subspace achieves such distance preservation and allows faster computation in $O(n^2 k)$ time, as opposed to $O(n^2 d)$ time.

Example 5.16 Given a collection of 64×64 pixel images (e.g., as in the MNIST dataset), where each pixel has 3 values specified by its R,G,B components, they can be represented as vector in 12,288D since $64 \times 64 \times 3 = 12,288$. Collections like these are common in data science for clustering problems or classification tasks (e.g., object recognition or identification in pictures). Such clustering algorithms usually compute the Euclidean distances between vectors, and each distance computation involves about 12,288 arithmetic operations, resulting in very slow performance. A much faster approach is to first randomly project the images to a much lower-dimensional space and then compute distances in that space.   □

This section describes a famous result for Euclidean spaces in functional analysis by Johnson and Lindenstrauss [14] that almost any projection to a lower-dimensional space achieves this effect, including some quantitative bounds on the dimension required. Specifically, the Johnson–Lindenstrauss (JL) lemma states that provided the target dimension k is sufficiently large, there is a projection from $\mathbf{R}^d$ to $\mathbf{R}^k$ such that *all* pairwise distances between a given set S of points are approximately preserved. How large a k is required? In fact, the proof of the lemma sketched below shows that such a projection can also be found easily in randomized polynomial time, as a projection to a random subspace of the required dimension.

Johnson Lindenstrauss Lemma
For every set S of n points in $\mathbf{R}^d$ and every $\varepsilon \leq 1/2$ and $k \geq 20 \log n / \varepsilon^2$, there exists a map $f : \mathbf{R}^d \to \mathbf{R}^k$ such that for every pair of points $u, v \in S$,

$$(1 - \varepsilon)\|u - v\| \leq \|f(u) - f(v)\| \leq (1 + \varepsilon)\|u - v\|.$$

There is also a generalization of this fact to manifold geodesic distances. Results by Baraniuk and Wakin [3] and improvements by Clarkson [4] show that a projection to a random subspace of the correct dimension (determined by the manifold) preserves geodesic distances as well. In this case, the mapped manifold is considered as a new manifold, and the geodesic distances are considered on that one. This result is intuitively clear as geodesic distances can be considered as path distances with a sufficiently dense sample on the manifold (as in ISOMAP). By the JL Lemma, the distances among the sample points are then preserved, and hence so are shortest path distances in the graph.

In practice, random projections are used quite often to do dimensionality reduction and visualizations as well.

Example 5.17 Figure 5.13 shows a random projection to $2D$ of the Swiss roll dataset.

Example 5.18 Figure 5.14 shows a random projection to $2D$ of the MNIST dataset. Since the images are binary and sparse and the projection is to a very low-dimensional space, the projection seems to have points divided into vertical and horizontal lines, i.e., many of the sparse high-dimensional images would project to the same first or second coordinate.

For the mathematically inclined reader, here is a sketch why the powerful JL lemma is true (simplified from the original by Frankl and Maehara [8], following Arriaga and Vempala [2], Dasgupta and Gupta [6], and Indyk and Motwani [13]. A proof that uses a sparser random matrix was also given by Achlioptas [1]). The basic reason why the JL lemma is true is a so-called norm preservation lemma.

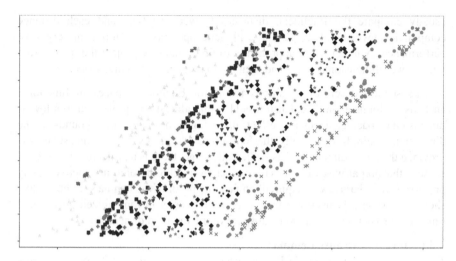

Fig. 5.13 The Swiss roll dataset projected to a random 2D subspace

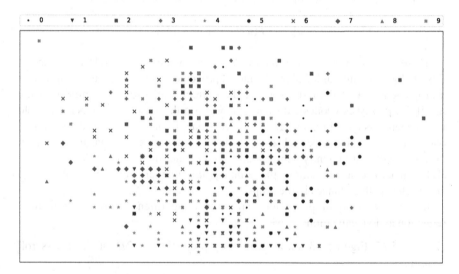

Fig. 5.14 The MNIST dataset projected to a random 2D subspace

Norm Preservation Lemma

Let $\varepsilon \in [0, 1]$ and $k \in \mathbf{N}$. If A is a random $k \times d$ matrix where each entry is a $N(0, 1)$ random variable, then for any $x \in \mathbf{R}^d$,

$$P\left((1 - \varepsilon)\|x\|^2 \le \|\frac{1}{\sqrt{k}}Ax\|^2 \le (1 + \varepsilon)\|x\|^2\right) \le 1 - 2e^{-(\varepsilon^2 - \varepsilon^3)k/4}.$$

The norm preservation lemma implies the JL lemma as follows. Basically, it suffices to find a distribution on random projections $f : \mathbf{R}^d \to \mathbf{R}^k$ for $k = 20 \log n/\varepsilon^2$ such that for any given $x \in \mathbf{R}^d$, the norm of x is approximately preserved under the mapping with high probability, $1 - 1/2n^2$. Since f is linear, it preserves norms approximately, and hence, it also preserves distances between points approximately since $d(u, v) = \|u - v\|$, and

$$d(f(u), f(v)) = \|f(u) - f(v)\| = \|f(u - v)\| \approx \|u - v\| = d(u, v),$$

where the first equality is by definition, the second by linearity, the third is implied by the norm preservation lemma, and the fourth again by definition.

To find such a projection, fixing any two points $u, v \in S$, the undesirable events E_{uv} happen when $\|f(u) - f(v)\|^2$ lies outside the interval $[(1 - \varepsilon)\|u - v\|^2, (1 + \varepsilon)\|u - v\|^2]$. By the norm preservation lemma above, there follows that $P(E_{uv}) \leq 2e^{-(\varepsilon^2 - \varepsilon^3)k/4}$. An estimate for this quantity for the range of ε can be given as

$$(\varepsilon^2 - \varepsilon^3)k/4 = \varepsilon^2(1 - \varepsilon)k/4 \geq \varepsilon^2(20 \log n)/8\varepsilon^2 > 2 \log n,$$

where the first equality is an identity, the second inequality uses $\varepsilon \leq 1/2$, and the third follows since $20 > 16$. Therefore, $P(E_{uv}) < 2e^{-2\log n} = 2/n^2$. Now, to approximately preserve all the interpoint distances in S, it suffices to show that none of the bad events should occur, i.e., $\bigcup_{u,v \in S} E_{uv}$ does not occur. The union bound implies that

$$P\left(\bigcup_{u,v \in S} E_{uv}\right) \leq \sum_{u,v \in S} P(E_{uv}) \leq n^2 \times 1/2n^2 = 1/2.$$

Thus the bad events happen with probability at most $1/2$. In particular, there must exist a sample point (i.e., a choice of the random A) such that the map defined by A preserves all interpoint distances in S approximately.

This argument is constructive because it provides clues to design a simple randomized algorithm to compute such an A. By simply sampling for an A by choosing kd independent Gaussians normally distributed as $N(0, 1)$, a good A can be had with probability at least $1/2$ since testing that A preserves all the interpoint distances in S is easy to do. If the first given choice fails, the procedure can be repeated. The probability estimate guarantees that by in worst case the second or third sample of A is good enough.

An outline of a proof of the norm preservation lemma uses advanced tools in probability theory that lie beyond the scope of this book. Nevertheless, here is a sketch for the technically minded readers. First, it uses an important property of the Gaussian distribution. If $g_1, \ldots, g_d$ be independent Gaussian $N(0, 1)$ random variables and $a_1, \ldots, a_d$ are arbitrary real numbers, then the random variable, $\sum_{i=1}^{d} a_i g_i$, is distributed as $\|a\|X$ where $X \sim N(0, 1)$, i.e., it is a Gaussian with mean 0 and variance $\|a\|^2$, where $a = (a_1, \ldots, a_d)^t$. This property of the Gaussian distribution is known as 2-*stability*. (The 2 indicates that it is the ℓ_2 norm of the vector a.) The theory of stable random variables is a deep topic in probability theory [7], and there are distributions that are α stable for any given α with $0 < \alpha \le 2$. The case $\alpha = 2$ is the Gaussian distribution, and $\alpha = 1$ corresponds to the Cauchy distribution. The general definition is that a linear combination of α stable independent random variables $a_1 X_1 + \ldots + a_d X_d$ will be distributed as $\|a\|_\alpha X$ where X has the same distribution as each X_i. (More details can be found at [7].) Second, row i of Ax is $[Ax]_i = \sum_{j=1}^d A_{ij} x_j$, and thus it is distributed as $\|x\| X_i$, where $X_i \sim N(0, 1)$. Moreover, all the rows are independent since the A_{ij} are all independent. Therefore, its expectation satisfies

$$E\left(\|\frac{1}{\sqrt{k}} Ax\|^2\right) = E\left(\frac{1}{k} \sum_{i=1}^k ([Ax]_i)^2\right) = E\left(\frac{1}{k} \|x\|^2 \sum_{i=1}^d X_i^2\right)$$

$$= \frac{\|x\|^2}{k} \sum_{i=1}^d E\left(X_i^2\right) = \|x\|^2.$$

Thus, while the variance of the squared norm $\|\frac{1}{\sqrt{k}} Ax\|^2$ is precisely $\|x\|^2$, it suffices to show that it is very close to $\|x\|^2$ with high probability. While the variance of $\frac{1}{k}\|Ax\|^2$ can be computed, a simple analysis of variance cannot give a result as strong as the norm preservation lemma. A strong concentration inequality [2] is required for the so-called χ^2 distribution (common distributions are summarized in Sect. 11.1) with k degrees of freedom. In other words, if a random variable X is the sum of squares of k independent Gaussians, i.e., $X = Z_1^2 + \ldots + Z_k^2$, where $Z_i \sim N(0, 1)$, and they are all independent, then $X \sim \chi^2(k)$ by definition. Now, the random variable $\frac{\|Ax\|^2}{\|x\|^2}$ is distributed precisely as $\chi^2(k)$. A lot is known about the $\chi^2(k)$ distributions including their density and distribution functions. Finally, the norm preservation lemma follows from the following lemma.

Concentration Lemma
If $X \sim \chi^2(k)$ is a random variable distributed as χ^2 and $\varepsilon \in [0, 1]$, then

$$P\left(X > (1+\varepsilon)k\right) \le e^{-(\varepsilon^2 - \varepsilon^3)k/4} \quad \text{and} \quad P\left(X < (1-\varepsilon)k\right) \le e^{-(\varepsilon^2 - \varepsilon^3)k/4}.$$

The proof of this lemma uses moment generating functions (MGFs), but it is a bit too involved to further pursue at this point. (More details can be found in [2].)

References

1. Achlioptas, D. (2003). Database-friendly random projections: Johnson-lindenstrauss with binary coins. *Journal of Computer and System Sciences, 66*(4), 671–687.

2. Arriaga, R. I., & Vempala, S. (2006). An algorithmic theory of learning: Robust concepts and random projection. *Machine Learning, 63*(2), 161–182.
3. Baraniuk, R. G., & Wakin, M. B. (2009). Random projections of smooth manifolds. *Foundations of Computational Mathematics, 9*, 51–77.
4. Clarkson, K. L. (2008). Tighter bounds for random projections of manifolds. In *Proceedings of the Twenty-Fourth Annual Symposium on Computational Geometry SCG'08* (pp. 39–48). New York: Association for Computing Machinery.
5. Cox, T. F., & Cox, M. A. A. (2000). *Multidimensional scaling* (2nd edn.). Boca Raton: Chapman & Hall CRC.
6. Dasgupta, S., & Gupta, A. (2003). An elementart proof of a theorem of Johnson and lindenstrauss. *Random Structures and Algorithms, 22*(1), 60–65.
7. Feller, W. (1971). *An introduction to probability theory and its applications* (vol. II, 2nd edn.). New York: Wiley.
8. Frankl, P., & Maehara, H. (1988). The Johnson-lindenstrauss lemma and the sphericity of some graphs. *Journal of Combinatorial Theory, Series B, 44*(3), 355–362.
9. Groenen, P. J. F., & Heiser, W. J. (1996). The tunneling method for global optimization in multidimensional scaling. *Psychometrika, 61*, 529–550.
10. Guttman, L. (1968). A general nonmetric technique for finding the smallest coordinate space for a configuration of points. *Psychometrika, 33*, 469–506.
11. Hinton, G. E., & Roweis, S. T. (2002). Stochastic neighbor embedding. In *Advances in neural information processing systems* (vol. 15, pp. 833–840). Cambridge: The MIT Press.
12. Hogg, R. V., McKean, J. W., & Craig, A. T. (2019). *Introduction to mathematical statistics* (8th edn.). London: Pearson.
13. Indyk, P., & Motwani, R. (1998). Approximate nearest neighbors: Towards removing the curse of dimensionality. In *Proc. of ACM STOC*.
14. Johnson, W. B., & Lindenstrauss, J. (1984). Extensions of lipschitz mappings into a hilbert space. *Contemporary Mathematics, 26*, 189–206.
15. Kumar, V., Grama, A., Gupta, A., & Karypis, G. (1994). *Introduction to parallel computing: Design and analysis of algorithms*. San Francisco: Benjamin-Cummings.
16. Lee, J. A., & Verleysen, M. (2007). *Nonlinear dimensionality reduction* (1st edn.). Berlin: Springer.
17. Maaten, L., & Hinton, G. E. (2008). Visualizing data using t-SNE. *Journal of Machine Learning Research, 9*(86), 2579–2605.
18. Messick, S. M., & Abelson, R. P. (1956). The additive constant problem in multidimensional scaling. *Pyschometrika, 21*, 1–15.
19. Nash, J. (1954). c^1 isometric imbeddings. *Annals of Mathematics, 60*(3), 383–396.
20. Shepard, R. N. (1962). The analysis of proximities: Multidimensional scaling with an unknown distance function. I. *Psychometrika, 27*(2), 219–246.
21. Silva, V., & Tenenbaum, J. B. (2002). Global versus local methods in nonlinear dimensionality reduction. In *Proceedings of the 15th International Conference on Neural Information Processing System, NIPS'02* (pp. 721–728).
22. Tenenbaum, J. B., Silva, V., & Langford, J. C. (2000). A global geometric framework for nonlinear dimensionality reduction. *Science, 290*(22), 2319–2323.
23. Torgerson, W. S. (1952). Multidimensional scaling: I. Theory and method. *Psychometrika, 17*(4), 401–419.

Chapter 6
Information-Theoretic Approaches

Max Garzon ⓘ**, Sambriddhi Mainali, and Kalidas Jana**

Abstract An entirely different but extremely relevant approach to dimensionality reduction can be taken using a different criterion, namely quantifying the information content of the features involved, within themselves or in relation to others. It turns out that Shannon's definition of information yields surprisingly interesting reductions. This chapter discusses five major variations of this idea, including comparisons using the concept of mutual information previously used in statistics and machine learning.

The problem of reliable telecommunication across a noisy channel (such as a phone line or extraterrestrial space between planets) led Shannon to fundamental research, an objective definition of *information* and the well-known theory of error-detecting and error-correcting codes in his foundational paper [1]. The theory blossomed into the field of information theory. The key concept in this field, Shannon entropy, quantifies the degree of *uncertainty* of (complementary to information in) a random process. This metric provides the mathematical foundation for information-theoretic analyses of channel capacity that characterize the maximum amount of information that can be transmitted through a noisy channel, while allowing noise removal without loss of information [1]. It has been also interpreted as a measure of the degree of *randomness* and/or *diversity* in a stochastic process. (The concept of entropy itself can be traced back to [2], later used in physics for heat theory and thermodynamics by [3], but here it will only refer to Shannon entropy and be just referred to as entropy and denoted by the customary H.)

Independence between features can be quantified using Shannon's conditional entropy $H(X_1 \mid X_2)$ between two features X_1 and X_2. When this entropy is low,

M. Garzon (✉) · S. Mainali
Computer Science, The University of Memphis, Memphis, TN, USA
e-mail: mgarzon@memphis.edu; smainali@memphis.edu

K. Jana
Fogelman College of Business, Memphis, TN, USA
e-mail: kjana@memphis.edu

M. Garzon et al. (eds.), *Dimensionality Reduction in Data Science*,
https://doi.org/10.1007/978-3-031-05371-9_6

X_2 essentially determines X_1, and thus X_1 is not an informative feature and could be discarded in favor of X_1. Five major variations of this kind will be reviewed.

The goal of this chapter is to show that the concept of entropy can be very effective for dimensionality reduction in unexpected ways.

6.1   Shannon Entropy (H)

This section provides definitions of Shannon entropy, conditional entropy, and their interpretations, along with a description of software libraries that can be used when manual computation becomes prohibitively costly, for example, when working with large datasets. Shannon entropy affords a nonlinear and nongeometric approach to dimensionality reduction based on information theory. A feature selection strategy can be based on the concept of conditional Shannon entropy of a random variable. (Sect. 11.1 defines background concepts in statistics and probability.)

According to Shannon, the *information content* $I(a)$ provided by an observation $[X = a]$ (or just $X = a$) of a random variable (RV) X on a probability space with a sample space Ω is the real number

$$I(a) = -\log_2 (P(X = a)),$$

where $P(X = a)$ is the probability of the event $[X = a] = \{e \in \Omega : X(e) = a\} = X^{-1}(a)$ associated with an observation of a value a for X.

The Shannon *entropy* $H(X)$ of a discrete RV X is the expected value (mean) I of the information content of the observations of all possible values of X, i.e., if X takes on only a finite number of values $a_1, \ldots, a_n$ with corresponding probabilities $P_1, \ldots, P_n$, then the entropy of X is given by

$$H(X) = -\sum_{i=1}^{n} P(X_i = a_i) \log_2 P(X_i = a_i) = -\sum_{i=1}^{n} P_i \log_2 P_i . \qquad (6.1)$$

Since it takes just about $\log_2(n)$ bits to express an integer n in, say, binary, $I(a)$ amounts to the average number of bits necessary to remove the *uncertainty* in answering a question like *what is the value of the observation* of X? when performing the random experiment in the background probability space. This is the key idea in Shannon's definition of *information* (content).

Entropy can thus be regarded as a measure of the average uncertainty in determining any given outcome of an observation of X or its quantification as the average number of bits necessary to identify all possible (unique) values of X (such as the 2 outcomes of a Bernoulli trial, somewhere between 0 and 1 bit, or the 6 outcomes of the roll of a die, somewhere between 2 and 3 bits). (If natural logarithms are used, the unit is called "nats" not bits. Throughout this book, entropies are reported in bits.)

Example 6.1 (Entropy of a Bernoulli Trial) In a Bernoulli trial (defined in Sect. 11.1), i.e., a sample space Ω with just two outcomes (*success* and *failure*, or heads and tails, or simply 1 and 0) with probabilities, p and $1 - p$, respectively, the RV X with value $X = 1$ if and only if the outcome is a success ($X = 0$ if it is a failure), the probability distribution of X is $P(X = 1) = P$ and $P(X = 0) = 1 - P$. The entropy is

$$H(X) = -P \log_2(P) - (1 - P) \log_2(1 - P).$$

□

Figure 6.1 shows the graph of the Shannon entropy $H(X)$ of the Bernoulli RV X. It is a concave function of P that attains a minimum value of zero for $P = 0$ and $P = 1$ and reaches a maximum value of 1 bit at $P = 0.5 = 1 - p$. Thus, the entropy is 0 when the outcome of the trial is a sure event (implying that there is no uncertainty in the outcomes of the random experiment), but the entropy is maximum when the outcomes are equally likely.

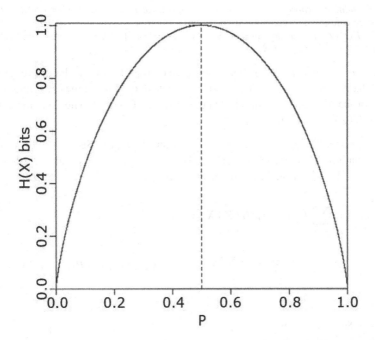

Fig. 6.1 For a Bernoulli random variable X with just two outcomes, the uncertainty is maximum $H(X) = 1$ when the two outcomes are equally likely (with probability $P = \frac{1}{2}$) and it is minimum when one of them is certain ($P = 1$, hence the other impossible $1 - P = 0$, or vice versa)

Example 6.2 (Entropy of a Roll of a Fair Die) If a die is fair, then a throw of the die
has 6 equally likely outcomes, each with probability 1/6. Therefore, the Shannon
entropy in this case is

$$H(X) = -\sum_{i=1}^{6} P_i \, \log_2 P_i = -\sum_{i=1}^{6} \frac{1}{6} \, \log_2(\frac{1}{6}) = \log_2 6 = 2.56 \text{ bits.}$$

□

There is more uncertainty in this stochastic process than in a Bernoulli trial
because 2.58 > 1, which makes sense intuitively because there are now 6 > 2
possibilities. In other words, knowing the outcome of a dice roll is more informative
(removes more uncertainty) than knowing the outcome of a coin toss indeed!

If two RVs X and Z on the same sample space are correlated (e.g., the value up
on one of the dice in a roll of two and the sum of the two values for both dice), it is to
be expected that knowing the value of the outcome of the corresponding experiment
(rolling the dice) for the observations for one of them will reduce uncertainty in the
value of the other. This reduction can be quantified precisely as follows:

The *conditional entropy* $H(Z|X)$ of a RV Z on (or *relative to*) another RV X is the
average entropy of Z conditioned on the observation of a value of X, i.e., the expected
value $E(H(Z_a))$ of the entropies of the RVs given by $Z_a : [Z \mid X = a]$, across all possible
values of a in the range of X.

Example 6.3 (Bivariate RVs) For conditional entropy for two scalar discrete RVs X
and Z, let X take on two values 1 and 2 and Z take on three values 1, 2, and 3, with
a joint probability distribution of X and Z given in Table 6.1. The Shannon entropy
of Z is $H(Z) = 1.56$ bits. □

Thus, given the joint probability distribution (defined in Sect. 11.1) of a sample
of observations of X and Z, where X takes values $x_1, \ldots, x_m$ and Z takes values
$z_1, \ldots, z_n$, the conditional entropy of Z given X is

$$H(Z|X) = \sum_{i=1}^{m} P(X = x_i) H(Z \mid X = x_i)$$

$$= -\sum_{i=1}^{m} P(X = x_i) \sum_{j=1}^{n} P(Z = z_j | X = x_i) \, \log_2 (P(Z = z_j | X = x_i)).$$

Table 6.1 Shannon entropy
of Z and conditional entropy
of Z given X for the joint
probability distribution in the
entries

$Z \setminus X$	1	2
1	0.15	0.10
2	0.20	0.20
3	0.30	0.05

$H(Z) = 1.56$ bits
$H(Z|X) = 1.48$ bits

This conditional entropy can be interpreted as the *uncertainty left in the values of Z given the observations of covariate feature X* in a given data point Z_{x_i}, averaged across all data points x_i.

Example 6.4 (Conditional Entropy) If the joint probability distribution is as given in Table 6.1, the conditional entropy of Z given X is $H(Z \mid X) = 1.48$ bits. □

The conditional entropy can be generalized to any number of conditions, i.e., $H(Z \mid X_1, X_2, \cdots, X_m)$ can be interpreted as the average uncertainty left in feature Z given the values of joint observations $X_1 = x_1, X_2 = x_2, \cdots, X_m = x_m$ in the same data points.

This definition of conditional entropy will be applied in Sect. 6.4 to achieve dimensionality reduction to 12 or 6 features in a dataset of 345 malware features taken from a sample of Microsoft's Malware Classification dataset with 1805 features (described in Chap. 1 and Sect. 11.4). Because the sample dataset is big, manual computation of conditional entropy is prohibitively costly. Software will come in handy to make computation manageable for such big datasets. One such software is the R package `infotheo`. It computes Shannon entropy using the Miller–Madow asymptotic bias corrected empirical estimator (which requires discretized features, and the option `equalfreq` can be used to discretize the data, where necessary, as illustrated in Sect. 6.4).

The Shannon entropy is also defined for continuous random variables as well. However, currently available computing software such as `infotheo` uses only discretized data even if it is continuous at the source. This is not really an issue since most data nowadays are collected with digital sensors and are thus discrete. Otherwise, truly continuous data has to be processed on conventional computers and so it has to be digitized in the process anyway.

6.2 Reduction by Conditional Entropy

This section and the next introduce alternative methods to use Shannon's entropy to reduce dimensionality in datasets with too many features, along with an assessment of their effectiveness in preserving significant information from the original dataset. The key idea here is to select more informative features or remove features whose information content is determined by the selected features, as determined by Shannon conditional entropy.

Example 6.5 The problem [**MalC**] of Malware Classification calls for an assignment of types to malwares from a predetermined set of types. It is a major problem in cyber security, where new malware is emerging at an alarming rate (for instance, more than 4.62 million new instances of malicious code were detected in June–July, 2019 [4].) A rigorous analysis to this problem is critical to study the evolution of malware and in developing appropriate countermeasures to contain cybercrime. Such an analysis can be either static or dynamic. A dynamic analysis depends on the execution of malware in a controlled environment [5, 6] and is costly effortwise [7]. On the other hand, a static analysis relying on decompilation tools (like IDA Pro) is more effective and efficient [8, 9]. However, it suffers from a major information retrieval issue since important information (like code layout, meta annotations, and even source language) in the source code is usually lost in the compilation process and cannot be retrieved in decompilation.   □

This section addresses the problem only indirectly by identifying important features to classify a piece of malware into categories for which known counter-measures are available. There are just too many features that can be associated with a piece of malware (e.g., 1809 in the Microsoft's dataset, described in Sect. 11.5.) Variants of entropy methods can be used to reduce the dimensionality of pre-identified features of these malwares to solve the classification problem. A first reduction by Ahmadi [10] (*arxiv.org/abs/1802.10135*) in 2016 used only 344 features extracted from the original Microsoft Malware Classification Challenge containing more than a thousand features [11].

The versatility and effectiveness of these methods can be illustrated with a second type of problem, the noisy classification problem (described in detail in Sect. 11.3.)

Example 6.6 One question of interest in any approach to DR is how good the approach is at identifying dependencies (statistical or other) in the features in the dataset. In a controlled experiment, a synthetic set can be designed with perfect knowledge of these dependencies from independent (e.g., randomly generated) raw (primitive) features. The effectiveness of DR methods can then be assessed by how well they discover these hidden, yet most relevant and independent features from a full set that includes other confounding features derived from the few primitive ones.

The primitive features were generated using the method described in [12] and publicly available as an API in Python at *sklearn.datasets.make_classification*. The primitive features are mixtures of several Gaussian clusters located near corners of the 12D hypercube and correspond to the labels in a classification problem. For each class label, the informative features are drawn independently from the standard normal distribution $N(0, 1)$. Four more dependent features were added as various linear combinations (with random coefficients) of the primitive features, as provided in the API. In the second phase, 6 more predictors were generated as repeats of two randomly selected (but uniform for all data points) primitive features/columns. One more feature was added as the sum of squares of two features selected randomly, one more feature consisting of the values of the predictions of a linear regression model fitted using two other randomly selected features as predictors, and the next feature was the deviation in the prediction from the true value. The last predictor was obtained in a similar manner but using the squares of the randomly chosen feature to predict one from the other. One last feature was generated as the outcome of the natural *logarithm* of a randomly selected but uniform predictor (a transformation that does not change entropy), for a total of 22 predictors. (A detailed description of how the synthetic datasets were generated is given with the datasets SYN12 and SYN23 in Sect. 11.5.)

A second dataset was generated likewise but halving the number of all parameters involved in generating features in the first set.   □

These datasets are used in Sect. 6.3 as well for assessment of the variant of the method being discussed there.

For this variant of entropy methods, it is important to quantify the informational independence between two predictors, unlike a dependent variable and a predictor used in conventional methods. Such quantification is done using Shannon conditional entropy between two predictors X_1 and X_2, i.e., $H(X_1 | X_2)$. When this entropy is high, knowledge of the values of X_2 removes little of the uncertainty in the values of X_1 and selection of X_1 may be necessary even if X_2 has been included [13]. Conversely, when this entropy is zero (or very low), X_2 (essentially, respectively) determines X_1, and thus X_2 is an informative feature and would suffice because models in a solution could derive any information in X_1 from it. This concept is analogous to the concept of multicolinearity in regression models described in Sect. 2.1. Multicolinearity is the condition that occurs if a few or all predictors are linearly correlated. As with multicolinearity, the information dependence between predictors is undesirable.

This section illustrates the application of two variants of this conditional entropy first introduced in [13]. A few significant predictors from the full list of predictors are obtained that possibly help in ascertaining target feature values (e.g., malware type.) To use as few Xs as possible, only predictors informative of the target are of interest. In particular, they should be as independent from other predictors as possible. Information-theoretically, that means that the conditional entropies relative to the other predictors (i.e., the uncertainty left in the predictor given another predictor's value) should be high overall. Thus, to decide whether to include X_i, the average is computed as

$$avg_i = avg \ \{H(X_i | X_j) : 1 \le j \ne i \le m\}$$

of the $H(X_i | X_j)$ over all other X_j to quantify how informative predictor X_i is as compared to others, for each i in the range of p predictors (excluding the target feature). For a given dataset, as many features as desired can thus be selected from the top features *maximizing* this average, after sorting the list in decreasing order.

A second paired variant of the same idea is to use the double conditional entropy to compute $H(X_i | X_j, X_k)$, for all j and k to determine how informative X_i is given the pair X_j, X_k when compared to all other pairs. If the average of these entropies is high, X_i should be selected, as above. The same process can be repeated to select more features as needed.

To compute conditional entropies between two predictors, `infotheo` package available in R was used, as mentioned above. (The package can be installed using the "install.packages(`infotheo`)" command in the R-console, as described in Sect. 11.6.) The Microsoft Malware Classification dataset was used to select two datasets using each variant. The first dataset contained only six predictors and the second contained 12 predictors. Then, these datasets were fed to several statistical and machine learning algorithms to assess the performance of these variants based on the performance of the resulting solutions for the classification problems on the datasets SYN12 and SYN23.

To assess the quality of this method of DR, one can proceed in two ways. First, the criterion being used (comparison of the information-theoretic content of the features) provides a good rationale why the choices may be effective and interpretable. However, there remains the issue of whether Shannon entropy really defines information as humans conceive of it, a much harder question that has remained unanswered. Alternatively, one can compare the effect of the choices on how good solution models are (as described in Sect. 2.4) when obtained on various sets of predictors. Thus, machine learning models trained on these reduced sets of predictors were compared against the scores yielded by machine learning models trained on the full set of predictor features, for both the Malware and synthetic datasets.

Following the standard procedure described above in Sect. 2.4, these datasets were split into training and testing subsets in a proportion of 80%–20%. The machine learning models included Linear Discriminant Analysis (LDA), Support Vector Machine (SVM), Multinomial Logistic Regression (MLR), Gaussian Naive-Bayes (GNB), Random Forest (RF), k-Nearest Neighbors (**kNNs**), and Neural Networks (NNs). The statistical models (the first four) were implemented using R scripts and the machine learning models were implemented using Python code (as in Sect. 11.6.) As usual, the F1-scores were selected as the metric to assess the quality of the solution models trained and these scores and are shown in Fig. 6.2 and Table 6.2 in the next section.

The methods have been used with similar success on other datasets, including [**BioTC**] and the synthetic datasets, selecting sets of various sizes (6 and 12 features) as well, as described in [13]. In terms of solutions, some models (RFs, kNNs) perform significantly better regardless of the problem ([**MalC**], [**BioTC**], or [**NoisC**]), while others (SVMs, LDAs, and MLRs) perform very inconsistently across these three representative problems. On average, solution models using only 6 features mostly performed poorly as compared to those using 12 features. Six (6) features might be too few for solving a complex problem, so reducing features too much may hurt for abiotic data.

Moreover, these DR methods offer the additional advantage of being computationally efficient because they are parallelizable. To compute conditional entropies between a pair of features (either two predictors or one predictor plus a target), the information about other features is not required, and hence several disjoint subsets of data can always be extracted and assigned to different computing nodes. This makes the process of feature extraction feasible by parallelization, even for big datasets.

In summary, conditional entropy performs competitively, if not very well, across the board compared to random selection of features or other methods.

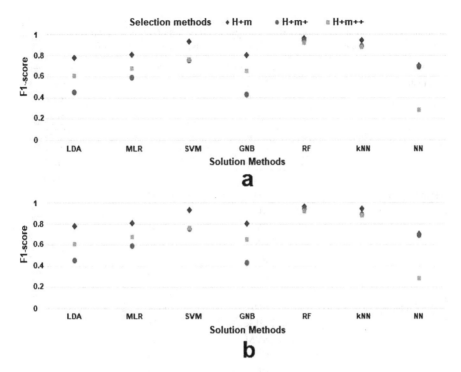

Fig. 6.2 Overall performance of DR to (**a**) 6 features (*top*) and (**b**) 12 features (*bottom*) based on conditional entropy on the problem of Malware Classification [**MalCP**]. Conditional entropy performs competitively, if not very well, with most machine learning solutions, The DR methods are selection by conditional entropy on predictors only (single H+m, paired H+m+, iterated H+m++) excluding targets, except LDA, GNB, and NN for paired iterative entropies (H++)

6.3   Reduction by Iterated Conditional Entropy

The results of Shannon conditional entropy for DR discussed in the previous section suggest the possibility that the interaction between various predictors at various levels might (perhaps jointly) contain better information about other predictors (which might have been deemed as most informative by themselves.) Therefore, another interesting variation is to select features based on a recursive procedure in which earlier choices affect together later selections.

In order to demonstrate the effectiveness of this alternative method for DR, the same procedure problems and datasets presented in the examples in Sect. 6.2 were used to assess the quality of machine learning models being trained using features selected by this alternative (indicated by H+m++). The scores are presented in Figs. 6.2 and 6.3, together with previously discussed variants of conditional entropy. On average, machine learning models seem to give better scores when the dimensionality reduction is done using the iterated conditional entropy method on the malware dataset. There seems to be no significant difference in the performance

Table 6.2 Performance comparison (F1-scores) between machine learning models trained using all features (*last column*) and those trained using reduced features (*second* and *third columns*) from the large synthetic dataset SYN23 for the problem of Noisy Classification [**NoisC**]

DR variant	Solution	Reduced	All features
All features	LDA		0.6937
	MLR		0.8013
	SVM		0.8690
	GNB		0.5085
	RF		0.9291
	kNN		0.9680
	NN		0.6860
H+m (Conditional)	LDA	0.9402	
	MLR	0.5728	
	SVM	0.8104	
	GNB	0.9096	
	RF	0.9717	
	kNN	0.9460	
	NN	0.8726	
H+m+ (Paired Conditional)	LDA	0.9484	
	MLR	0.9345	
	SVM	0.9902	
	GNB	0.9425	
	RF	0.9800	
	kNN	0.9578	
	NN	0.9082	
H+m++ (Iterated)	LDA	0.9481	
	MLR	0.9345	
	SVM	0.9906	
	GNB	0.9547	
	RF	0.9794	
	kNN	0.9910	
	NN	0.9417	

of machine learning models trained on the features selected using single and paired conditional entropies. Moreover, these models seem to give better performance when 12 reduced features are used as predictors instead of 6. On the other hand, the running times on an HPC of the relative entropy calculations to select features are in the order of minutes (single entropies), under 2 h (paired entropies) and under 6 h (iterated entropies.) So, there is a trade-off between the performance of the machine learning models and the computational resources required to select features using these information-theoretic DR methods.

In order to demonstrate that feature selection using entropy methods without target is most likely an optimal choice than using the whole dataset, the methods was tried also on the synthetic datasets mentioned above (described in Sect. 11.5).

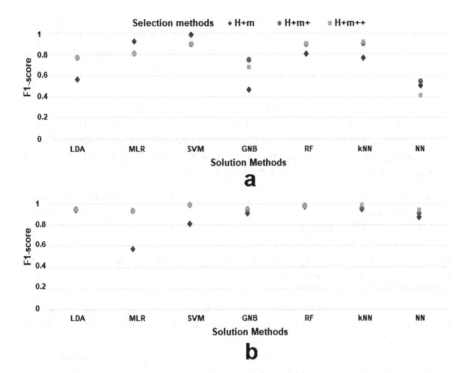

Fig. 6.3 Overall performance of DR to (**a**) 6 features (*top*) and (**b**) 12 features (*bottom*) based on conditional entropy on the problem of Synthetic Data Classification [**NoisC**] (with controlled dependencies) on synthetic datasets SYN12 and SYN23. Conditional entropy performs competitively, if not very well, with most machine learning solutions. The DR methods are selection by conditional entropy on predictors only (H+m, paired H+m+, iterated H+m++) excluding targets, except LDA, GNB, and NN for paired iterative entropies (H++)

Standard metrics like accuracy, precision, recall, and F1-score were used to evaluate the performance of different ML models trained and validated on these datasets. A dimensionality reduction method was deemed to be of a good quality if an average performance score of several machine learning models trained to solve these problems is above 81% (average of the scores reported in [13]) for both problems.

A comparison of the scores is shown in Table 6.2. As described above, the synthetic dataset was designed so as to contain some predictors that could be derived using some kind of combination of other predictors. Therefore, these features are not informationally independent. Although the impact of this dependency is not so evident in some machine learning models on the given dataset, the usage of all features led to solution models with the performance being dominated by those trained using reduced features on average. Therefore, as a general rule of thumb, one can easily conclude that in the presence of too many predictors, it is always a wise choice to look out for some dimensionality reduction methods to obtain only informationally rich features. However, there is still a question that remains unanswered, i.e., what is the threshold determining too many predictors? The answer to this question really ultimately depends on the type of data science problem at hand and the choice of features in the dataset. One can only hope to consider all constraints impacting the search for a solution, for example, answering the following questions (perhaps among others):

- Is it computationally feasible to include all features available?
- Is there enough time to train a model with all features and validate it?
- Are there enough datapoints to include all features (for example, a dataset containing only 10 points might be enough to train a statistical model if there is only one predictor, but not when more than six predictors are to be considered)?

As with conditional entropy in Sect. 6.2, the methods have been used with similar success on other datasets, including one for BioTC and the synthetic datasets SYN12 and SYN23, selecting sets of various sizes (6 and 12 features) as well, as first described in [13]. In terms of solutions, some models (RFs, kNNs) perform better regardless of the problem (**[MalC]**, **[BioTC]**, or **[NoisC]**), while others (SVMs, LDAs, and MLRs) perform very inconsistently across these three representative problems. On average, solution models using only 6 features performed mostly poorly as compared to those using 12 features. Six (6) features might be too few for solving a complex problem, so reducing features too much may hurt for abiotic data.

In terms of the computational efficiency, the advantage of being parallelizable is not as impressive. Because of their nature, to compute iterated conditional entropies between a pair of features (either both predictors or one predictor plus another target), data from other features is now required and the number of possible combinations is explosive. These facts make the process of feature extraction feasible less attractive, particularly for big datasets. Perhaps, a combination of the two methods, first selecting a smaller subset using conditional entropy and then using iterated conditional entropy of the smaller subset, may be a more productive approach.

In summary, conditional entropy and iterated conditional entropy perform competitively, if not very well, across the board compared to random selection of features or other methods.

6.4 Reduction by Conditional Entropy on Targets

This section shows how dimensionality reduction of predictor features can be achieved using conditional entropy of *the target feature* relative to the predictors. Two examples are used to illustrate reductions to a set of 6 or 12 features out of 344 singles features and 6 or 12 paired features in a sample dataset of 345 features from the Microsoft's Malware Classification dataset of 1805 features for predicting the target feature "Class" (type of malware). As a result, a principled argument can be made in Sect. 6.5 to show how selecting features by minimizing conditional entropy is equivalent to selecting features by maximizing mutual information with respect to the target feature, the commonly used approach in the literature on information-theoretic methods for feature selection.

In statistics, when selecting features for predictors, the target feature is naturally taken into account. In this variant, the degree with which a target Y is predictable given a feature X_i is decided using conditional entropy $H(Y \mid X_i)$ between Y on a feature X_i. When this entropy is low (high), the uncertainty is low (high) for the values of Y given the values of X_i, so X_i is informative (or is not, respectively.) Hence, X_i should be selected if this entropy is low. This is the opposite of the criterion in Sects. 6.2 and 6.3 when predictors are being compared for selection, but the approach is implemented just the same way, with the following changes:

- Calculate the $H(Y \mid X_i)$ over all X_i to quantify how informative predictor feature X_i is compared to others.
- Select the top features minimizing this entropy after sorting the list in increasing order.

As mentioned in Sect. 6.1, the conditional entropy $H(Z \mid X)$ can be generalized to any number of conditions $X_1, X_2, \cdots, X_m$ instead of a single condition X (Table 6.3).

Example 6.7 Table 6.4 shows 12 predictor features selected based on

$$H(Class \mid X_1, X_2, \cdots, X_m)$$

Table 6.3 Top 12 *single* features X_i $(i = 1, \cdots, 12)$ (*second column*) selected by sorting 344 values of conditional entropies $H(Class \mid X_i)$ in a sample dataset of 345 features taken from the Microsoft Malware Classification dataset of 1805 features for predicting the target feature "Class." Table 6.4 shows the values of the entropies

ID	Selected feature
1	db0_por
2	dbN0_por
3	db3_all
4	db3_rdata
5	dc_por
6	ent_p_1
7	ent_p_2
8	ent_p_4
9	ent_q_diff_diffs_10
10	ent_q_diff_diffs_11
11	ent_q_diff_diffs_1_mean
12	ent_q_diff_diffs_2_mean
13	ent_q_diff_diffs_2_min
14	ent_q_diffs_0
15	ent_q_diffs_mean
16	ent_q_diffs_var
17	known_Sections_por

Table 6.4 Top single conditional entropies $H(Class \mid X_i)$ (entries in a row i in the second column) and top conditional entropies $H(Class \mid X_i, X_j)$ (if more than one entry in a row in columns $j \geq 1$) of the top 12 *pairs* (X_i, X_j) obtained by sorting $\binom{344}{2} = 58{,}996$ values of $H(Class \mid X_i, X_j)$ corresponding to 58,996 pairs of the features in Table 6.3 used to solve the Malware classification problem [**MalC**]. A blank entry indicates that the corresponding feature was not selected, either as a single i or as part of a pair i, j

$i : j$	$H(Class \mid X_i)$	1	2	3	4	5	$\cdots$	17
1								
2								
3								
4								
5	1.756							
6	1.765							
7	1.756							
8	1.756							
9	1.521							0.898
10	1.548	0.913	0.916	0.882	0.880	0.878		0.889
11	1.747							
12	1.745							
13	1.681							
14	1.676							
15	1.566				0.908	0.874		0.875
16	1.586		0.915	0.909				
17								

with $m = 1$ and 2, respectively, for predicting the target "Class", i.e., the 345th feature in the sample dataset of 345 features taken from the population of Microsoft's Malware Classification dataset of 1805 features, by implementing the above two-step procedure using infotheo. It is worth mentioning that the number 12 should not be interpreted as any sort of "optimal" number of features in the sense of being determined by some optimality criterion. Rather, it is just a low number chosen for the purpose of comparison with choosing fewer features.   □

To assess the quality of this reduction, various machines learning solutions were computed for the Microsoft's malware dataset for the [**MalC**] problem, similar to the procedure for the previous variants of the entropy method. Table 6.5 shows the results. Furthermore, various machines learning solutions were also computed on synthetic datasets SYN12 and SYN23 (described in Sect. 11.5) with primitive features hidden from the conditional entropy reduction method. Tables 6.6 and 6.7 show the F1-scores for comparison.

In summary, a careful comparison with the other conditional entropy methods shows that, perhaps surprisingly, *taking into account the targets does not really seem to make such a significant difference* in the quality of the solutions to a data science problem. However, choosing 12 features rather than 6 again does help significantly

Table 6.5 Performance (F1-scores) comparison between machine learning models trained using 6 and 12 singles predictor features and 6 and 12 paired predictor features, selected by conditional entropy of the target feature "Class" in the Malware Classification dataset. (Conditional entropy of singles and paired features are denoted by H and $H+$, respectively)

Entropy variant	ML solution model	6 features	12 features	All features
All features	LDA			
	MLR			0.6934
	SVM			0.3772
	GNB			0.7013
	RF			0.8338
	kNN			
	NN			
	Adaboost			0.7802
H	LDA	0.3579	0.4624	
	MLR	0.6433	0.8398	
	SVM	0.7618	0.8152	
	GNB	0.3189	0.5871	
	RF	0.9746	0.9869	
	kNN	0.9593	0.9690	
	NN	0.6949	0.8397	
H+	LDA	0.7087	0.7946	
	MLR	0.8230	0.9217	
	SVM	0.8398	0.9223	
	GNB	0.1072	0.6151	
	RF	0.9883	0.9895	
	kNN	0.9125	0.9323	
	NN	0.7447	0.9584	

improve the quality of the solutions. Pairing now, however, does seem to make a difference as well, in general.

6.5 Other Variations

A common information-theoretic approach used in statistics and data science, in particular for feature selection, makes use of the concept of *mutual information* $I(Y : X)$.

The *mutual information* (also known as *information gain*) is given by

$$I(Y : X) = H(Y) - H(Y \mid X).$$

This concept rather quantifies information independence between two variables, analogously to the concept of statistical independence.

Table 6.6 Performance (F1-scores) comparison between machine learning models trained using 6 and 12 singles predictor features, and 6 and 12 paired predictor features, selected by conditional entropy of the target feature "Class" in the Noisy Classification dataset SYN13. (Conditional entropy of singles and paired features are denoted by H and $H+$, respectively)

Entropy variant	ML solution model	6 features	12 features	All features
All features	[14]			0.7970
	[15]			0.9600
H	LDA	1	1	
	MLR	0.9805	0.9828	
	SVM	0.9912	0.9946	
	GNB	0.8598	0.9358	
	RF	0.9806	0.9865	
	kNN	0.9840	0.9709	
	NN	0.9153	0.9826	
H+	LDA	1	1	
	MLR	0.9811	0.9830	
	SVM	0.9922	0.9946	
	GNB	0.8598	0.9358	
	RF	0.9809	0.9867	
	kNN	0.9840	0.9709	
	NN	0.9232	0.9807	

Table 6.7 Performance (F1-scores) comparison between machine learning models trained using 6 and 12 singles predictor features, and 6 and 12 pairs predictor features, selected by conditional entropy of the target feature "Class" in Noisy Classification dataset (SYN22). (Conditional entropy of singles or paired features is denoted by H or $H+$, respectively)

Entropy variant	ML solution model	6 features	12 features	All features
All features	[14]			0.7970
	[15]			0.9600
H	LDA	0.5635	0.9402	
	MLR	0.9246	0.5728	
	SVM	0.9865	0.8104	
	GNB	0.4654	0.9096	
	RF	0.8063	0.9717	
	kNN	0.7657	0.946	
	NN	0.5029	0.8726	
H+	LDA	0.9975	1	
	MLR	0.9292	0.9849	
	SVM	0.9875	0.9783	
	GNB	0.9940	1	
	RF	0.9910	9895	
	kNN	0.7646	0.9511	
	NN	0.7033	0.9779	

Example 6.8 The mutual information in the variables Z and X in Example 6.3 in Sect. 6.1 above is

$$I(Z:X) = H(Z) - H(Y \mid X) = 1.56 - 1.48 = 0.08 \text{ bits .}$$

□

Example 6.9 In the [**MalC**] problem, there are quite a few predictors in the original dataset (1809 to be exact). Even after using the reduced dataset by [10] (344 predictors), there are still too many, and they may hide a number of dependencies (e.g., some of these features might be collinear). One could alternatively follow the approach discussed in previous sections, but using the mutual information as a selection criterion instead of conditional entropy. □

A comparison of the definitions of conditional entropy and mutual information makes it is clear that they are complementary quantities in the entropy $H(Y)$ of the target Y, so that the lowest conditional entropy corresponds to the highest mutual information with a predictor X, and vice versa. Therefore, selecting features by maximizing mutual information, for any problem and dataset, is equivalent to selecting features by minimizing conditional entropy, as has been done in Sects. 6.1–6.4.

References

1. Shannon, C. E. (1948). A note on the concept of entropy. *Bell System Technical Journal, 27*(3), 379–423.
2. Clausius, R. (1879). *The mechanical theory of heat, nine memoirs on the development of concept of "entropy"*. McMillan and Co.
3. Boltzmann, L. (1878). On some problems of the mechanical theory of heat. *The London, Edinburgh, and Dublin Philosophical Magazine and Journal of Science, 6*(36), 236–237.
4. Yang, P., Zhou, H., Zhu, Y., Liu, L., & Zhang, L. (2020). Malware classification based on shallow neural network. *Future Internet, 12*(12), 219.
5. Fattor, A., Lanzi, A., Balzarotti, D., & Kirda, E. (2015). Hypervisor-based malware protection with access miner. *Computers & Security, 52*, 33–50.
6. Hashemi, H., Azmoodeh, A., Hamzeh, A., & Hashemi, S. (2017). Graph embedding as a new approach for unknown malware detection. *Journal of Computer Virology and Hacking Techniques, 13*(3), 153–166.
7. Pektaş, A., & Acarman, T. (2017). Classification of malware families based on runtime behaviors. *Journal of Information Security and Applications, 37*, 91–100.
8. Fan, C. I., Hsiao, H. W., Chou, C. H., & Tseng, Y. F. (2015). Malware detection systems based on API log data mining. In *2015 IEEE 39th Annual Computer Software and Applications Conference* (Vol. 3, pp. 255–260). IEEE.
9. Santos, I., Brezo, F., Ugarte-Pedrero, X., & Bringas P. G. (2013). Opcode sequences as representation of executables for data-mining-based unknown malware detection. *Information Sciences, 231*, 64–82.
10. Ahmadi, M., Ulyanov, D., Semenov, S., Trofimov, M., & Giacinto, G. (2016). Novel feature extraction, selection and fusion for effective malware family classification. In *Proceedings of the Sixth ACM Conference on Data and Application Security and Privacy* (pp. 183–194).

11. Ronen, R., Radu, M., Feuerstein, C., Yom-Tov, E., & Ahmadi, M. (2018). Microsoft malware classification challenge. *CoRR*, abs/1802.10135.
12. Guyon, I. (2003). Design of experiments of the nips 2003 variable selection benchmark. In *NIPS 2003 Workshop on Feature Extraction and Feature Selection* (Vol. 253).
13. Mainali, S., Garzon, M., Venugopal, D., Jana, K., Yang, C. C., Kumar, N., Bowman, D., & Deng, L. Y. (2021). An information-theoretic approach to dimensionality reduction in data science. *International Journal of Data Science and Analytics*, *12*, 185–203. https://doi.org/10.1007/s41060-021-00272-2
14. Donoho D. L. (2000). High-dimensional data analysis: The curses and blessings of dimensionality. In *AMS Conference on Math Challenges of the 21st Century*.
15. Achlioptas, D. (2003). Database-friendly random projections: Johnson-Lindenstrauss with binary coins. *Journal of Computer and System Sciences*, *66*(4), 671–687.

Chapter 7
Molecular Computing Approaches

Max Garzon ⓘ and Sambriddhi Mainali

Abstract Molecular approaches exploit structural properties built deep into DNA by millions of years of evolution on Earth to code and/or extract some significant features from raw datasets for the purpose of extreme dimensionality reduction and solution efficiency. After describing the deep structure, it is leveraged to render several variations of this theme. They can be used obviously with genomic data, but perhaps surprisingly, with ordinary abiotic data just as well. Two major families of techniques of this kind are reviewed, namely genomic and pmeric coordinate systems for dimensionality reduction and data analysis.

Molecular approaches deal with the use of properties of deoxyribonucleic acid (DNA) to code and extract some information-rich features from arbitrary raw datasets (including digital abiotic data) for extreme dimensionality reduction and solution efficiency of data science problems.

Advances in Internet technologies and sensing have enabled us to generate and/or store huge amounts of data (in the scale of exabytes per day), but we still lack methods to analyze them at the same scale and speed. A major roadblock is that most real-life data, such as text and images, are unstructured in the sense that they are not generated by a well-defined model or even organized in tabular format. This lack of structure makes it harder to select or extract useful critical information to solve problems at a humanly meaningful (semantic) level. Deep learning methods have handled these difficulties to an extent, but it is very difficult to explain how these methods learn significant features for such data analytics. DNA, on the other hand, has been processing humongous amount of biological information in the nick of time, as demonstrated by all living organisms in this planet (e.g., protein synthesis and self-organization) [1]. In particular, recent works have demonstrated that DNA deeply encodes enough information about an organism so that features about taxonomic group [2], phenotype [3], and, perhaps surprisingly,

M. Garzon (✉) · S. Mainali
Computer Science, The University of Memphis, Memphis, TN, USA
e-mail: mgarzon@memphis.edu; smainali@memphis.edu

145

even environmental conditions of the natural habitat where an organism grew up [4] could be predicted. This chapter demonstrates a novel approach to leveraging this property of DNA to solve some of these challenging problems.

Leveraging DNA's exquisite discrimination ability to solve computational problems is not a new idea. Adleman [5] pioneered the field of DNA computing by proposing to build computers using real DNA molecules. Eventually, it was realized that fundamental problems (such as CODEWORD DESIGN (**[CWD]**) below) would need to be solved to get DNA molecules to do something they did not evolve for better than electronic computers. These problems are in the class **NP**-complete (defined in Sect. 11.3) using any single reasonable metric that approximates the strength of double helix bonding, thus practically excluding the possibility of finding any procedure to find maximal sets exactly and efficiently [6]. The field then refocused on a potentially more impactful application, namely the self-assembly of complex nanostructures [7]. Rather than pursuing this line of research, this chapter is rather focused more on using DNA and processes in vivo (within living organisms) as an inspiration to develop new tools to solve problems in computer science and biology in silico (in simulations on conventional computers).

[CWD]   CODEWORD DESIGN

INSTANCE: A positive integer m and a threshold $\tau > 0$

QUESTION: What is a largest set B of single DNA strands of length m that do not crosshybridize to themselves or to their WC complements (an nxh set) under stringency τ, i.e., their hybridization distance $|xy| > \tau$ for all $x, y \in B$?

To demonstrate the efficiency of these dimensionality reduction methods, several fundamental and challenging problems in biology and computer science were selected in [2, 3], and [4], i.e., species identification **[BioTC]** (or BioTaxonomy as referred to in this chapter), phenotype prediction **[RossP]**, and habitat prediction **[LocP]** (all defined in Sect. 11.5). Species identification calls for a solution model to identify a species of an organism from a predefined set of species given its DNA sequence. All three kinds of problems can be solved using the same techniques, so only results for **[BioTC]** are described in detail, in addition to the running examples, such as **[MalC]** and **[NoisC]** (also described in Sect. 11.5) for abiotic data.

[BioTC]   BIOTAXONOMIC CLASSIFICATION (T)

INSTANCE: A (long) DNA sequence (over the alphabet $\{a, c, g, t\}$) from a living organism x

QUESTION: What species in T does x belong in?

The features representing an organism could be, for example, mitochondrial genes COI, COII, COIII, and CytB from the organism's genome. Other choices give rise to different problems, as shown in the corresponding datasets in Sect. 11.5, e.g., a most complex problem will arise if the full genome is used to represent an instance of a biological organism.

In biology, **[BioTC]** is a classical problem first formulated by Carl Linnaeus' idea of a catalog of life on Earth into a hierarchical clustering system [8] back in the 1700s. The literature available showcasing the attempts of biologists in solving this problem reflect a slow shift, from considering only morphological features, to relying only on molecular data (e.g., their genomic sequences), to a very recent integrative approach of considering their genome, morphology, behavior, geographic distribution, ecology, and so on to make such distinction [9, 10]. However, the scope of this chapter is only focused on the more challenging problem of using the genomic sequences of organisms *alone* for data analysis since DNA may encode enough information to derive some of these features as well.

To assess the performance of solution models trained using the features extracted using biomolecular methods, two datasets are used, namely, the Malware dataset described above and a biotic dataset of about 249 biological organisms spread across 21 species constructed for this purpose. These species and their representation are shown in Table 7.1. Four significant genes from the mitochondrial DNA of these organisms were chosen, namely three subunits of *Cytochrome c oxidase* and *Cytochrome b*. The corresponding DNA sequences were downloaded from GenBank [11].

Table 7.1 Organisms in the sample data for biotaxonomy classification **[BioTC]** of a biological taxon. The 21 classes/labels in the partition are in the first column (the dataset is fully described in Sect. 11.5)

Label	T: Genus species	Common name	Count
1	*Apis mellifera*	Western honeybee	4
2	*Arabidopsis thaliana*	Thale cress	5
3	*Bacillus subtilis*	Hay/grass bacillus	18
4	*Branchiostoma floridae*	Florida lancelet	18
5	*Caenorhabditis elegans*	Round worm	6
6	*Canis lupus*	Wolf	18
7	*Cavia porcellus*	Pork	4
8	*Danio rerio*	Zebra fish	9
9	*Drosophila melanogaster*	Fruit fly	18
10	*Gallus gallus*	Red junglefowl	18
11	*Heterocephalus glaber*	Naked mole rat	3
12	*Homo sapiens*	Human	18
13	*Macaca mulatta*	Rhesus macaque	8
14	*Mus musculus*	House mouse	18
15	*Neurospora crassa*	Red bread mold	5
16	*Oryza sativa*	Asian rice	12
17	*Pseudomonas fluorescens*	Infectious bacterium	6
18	*Rattus norvegicus*	Brown rat	18
19	*Rickettsia rickettsii*	Tick-born bacterium	18
20	*Saccharomyces cerevisiae*	Yeast	18
21	*Zea mays*	Corn/maize	7
			249

7.1 Encoding Abiotic Data into DNA

This section describes how to encode abiotic (e.g., malware) data into DNA sequences for problem solving in data science.

The idea of leveraging DNA's exquisite discrimination ability to solve computational problems was introduced in the 1990s [5]. Several attempts have been made in that direction for abiotic data [12–14]. The authors in [14] used solutions to [**CWD**], namely encoding data into DNA using so-called noncrosshybridizing (nxh) bases, to solve Word Disambiguation and Textual Entailment in semantic analysis. However, to be able to process abiotic data (e.g., text and images) using DNA, an automated method to encode and transform abiotic data into DNA sequences must be used.

Example 7.1 The obvious and straightforward approach is to make homomorphic substitutions of alphabet symbols, for example, substitute characters in English text with their ASCII codes in binary and then convert binary strings into DNA sequences by concatenating encodings of their characters in the same order. For example, [15] simply uses DNA codons (DNA sequences of length 3 coding for proteins) to homomorphically encode individual characters in computer files totaling 739 kilobytes of hard disk storage encoding all 154 of Shakespeare's sonnets using a cipher where every byte of 0s and 1s is represented by a word using 4 symbols a, c, g, and t. This method appears to be too coarse and noise prone to hybridization patterns that may overload a model [13]. On the other hand, using too sophisticated encoding methods that hide some processing might dilute the credibility of a model to provide better solutions. Therefore, a straightforward encoding of abiotic data is discussed next to demonstrate the ability of DNA to process abiotic data, although many other methods are certainly possible.   □

A piece of malware can be regarded as an ordered set of hexcodes representing a string of hexadecimal characters. Thus, a DNA encoding can be simply obtained by converting each hexadecimal character to its binary equivalent and then concatenating these binaries into a sequence in DNA form. Two bits encode four possible strings that can be regarded as a, c, g, or t, e.g.,

$$00 \rightarrow a; \quad 01 \rightarrow c; \quad 10 \rightarrow g; \quad 11 \rightarrow t.$$

A subset of 511 malwares from all malwares in the Malware Challenge dataset was drawn, and they were encoded into DNA sequences using this procedure.

In order to show the results, the method to reduce dimensionality using DNA requires some other concepts in the following section.

7.2 Deep Structure of DNA Spaces

It is well known that, as the blueprint of life, DNA encodes for critical information required to develop and sustain life in every living organism (e.g., protein synthesis and self-organization) due to its self-organizing properties. This critical role of DNA motivated work in a new field, DNA computing, inspired by the ideas of using DNA itself as a computational medium pioneered by Adleman [5] and as smart glue for self-assembly applications by Seeman [7] and Winfree [16]. Eventually, it was realized that fundamental problems (such as CODEWORD DESIGN [CWD] below) would need to be solved to get DNA molecules to do something they did not evolve for better than electronic computers. Finding a solution to this problem is **NP**-complete using any single reasonable metric that approximates the Gibbs energy, thus practically excluding the possibility of finding any procedure to find maximal sets exactly and efficiently [6]. The field then refocused on a potentially more impactful application, namely the self-assembly of complex nanostructures [7] in vivo (in living organisms) and, primarily, in vitro (test tubes).

A parallel effort continues as well to use DNA and hybridization as an inspiration to develop new tools to solve problems in computer science and biology in silico i.e., in simulations on conventional computers. In particular, recent works further demonstrate that DNA sequences encode enough information about an organism so that features about phenotype, taxonomic group, environmental conditions of the natural habitat where an organism lived, and so on could be predicted [2–4]. The most fundamental and powerful property of DNA is hybridization, its *exquisite discriminating ability in forming double strands* (helices) as discovered by Watson and Crick [1]. A systematic attempt to develop tools to tackle this challenging **[CWD]** has been steadily pursued by several authors, e.g. [6, 17–20], because of its importance, both fundamentally and bioinformatically. Useful insights into the structure of Gibbs energy of DNA duplex formation governing hybridization have been revealed through a *metric* approximation of the Gibbs energy between two DNA oligonucleotides of the same length, known as the h-distance. This distance allows topological sorting of DNA oligonucleotides in DNA spaces in such a way that the physical distance between any two points in the space reflects their hybridization affinity. In particular, two Watson–Crick (WC) complementary oligonucleotides (e.g., aaa and ttt for oligonucleotides of length 3) are at distance 0 and collapse into a single point (the pair aaa/ttt, likewise for acg/cgt). Such pairs are referred to as paired mers or simply pmers. Furthermore, this distance allows us to reason about hybridization using geometric and spatial analogies with the concepts of ordinary distance that people use so easily to facilitate reasoning and conceptualization of the world, as described in Sect. 1.2. For example, aaa/ttt (ccc/ggg) is located as sort of a north (south, respectively) pole in the DNA space of 3-pmers, as shown in Fig. 7.2a. These concepts enable understanding of the complex Gibbs energy landscapes of DNA hybridization that are fundamental to genomics and physiology in living organisms. In particular, they provide important clues how to strategically select sets of pmers for probes in a microarray design that would optimize not only the extraction of information deeply encoded in the genomes of living organisms but also find hidden structure in apparently unstructured abiotic data codes into DNA sequences, as illustrated next.

Example 7.2 Here are some questions that DNA might know something about:

- Does DNA encode enough information to enable *quantitative predictions* of phenotypic features in a biological organism, even though they depend on environmental factors presumably beyond DNA?
- Does DNA actually encode enough information to say something informative about environmental factors (e.g., latitude, longitude, temperature, and so on) of the natural habitat where these organisms grew and lived?
- Is a principled and taxon-independent definition of the concept of *species* possible that is universal for biological taxonomies?
- Can we provide an objective taxon independent and systematic definition of pathogenicity shared between hosts (e.g., *homo sapiens*) and micro-organisms (e.g., bacteria and fungi) based on a computational approach?

$$\square$$

In order to make inroads into these questions, one must take a very deep look into DNA sequences from the point of view of computer science. This inquiry will yield impressive reductions in the dimensionality of DNA data and even abiotic sequences below to address the problem of what kinds of information are encoded in DNA.

7.2.1 Structural Properties of DNA Spaces

This section describes a number of structural properties of DNA spaces. Some precise definitions of concepts and terms are required to elucidate and describe them.

Given a positive integer m, a *DNA sequence* x is a string defined over the alphabet $\Sigma = \{a, c, g, t\}$. They will also be referred to as m-mers, where $m = |x|$ is the length of string x. The *Watson–Crick (WC) complement* of x is the string x' obtained after first taking the reverse of x (i.e., x^r) and replacing every a (c) by t (g, respectively) and vice versa, i.e., $x' = (x^r)^c$. A *pmer* (or $|x|$-pmer) is an (unordered) pair of two WC complementary DNA sequences $\{x, WC(x)\}$ (simply denoted x/x', or just x, the lexicographically first element in the pair). The *DNA space* $\mathbf{D}_m$ of length $m > 1$ consists of the set of all such m-pmers.

If x is a WC-palindrome (i.e., $WC(x) = x' = x$), then the corresponding pmer is really a single string. For reasons that will become apparent below, WC-palindromes will be excluded from consideration throughout.

Hybridization is governed by the familiar Gibbs energy, the chemical equivalent of the potential energy in physics, which depends on physical parameters (such as the internal energy, pressure, volume, temperature, and entropy) of the environment in which the duplex is formed. The more negative the Gibbs energy, the more stable the duplex formed. Unfortunately, the available models in biochemistry provide no gold standard to assess Gibbs energies other than just accepted empirical approximations [21]. The most popular method to approximate the Gibbs energy is the so-called Nearest Neighbor (NN) Model, but this model lacks all the three properties of a *metric* (defined in Sect. 1.2) for a good approximation that lends itself to deeper analysis. Furthermore, the size and composition of nxh sets are very difficult to establish in this model due to the lack of intuition and tools as to the structure of the Gibbs energy landscapes [6].

In the field of DNA computing, many attempts have been made to address this issue. They have revolved about the [CWD] problem identified as a fundamental problem in the field. Adleman [5] emphasized the need of a good coding strategy for using DNA to process information. A coding scheme is crucial to experiments in vitro, for mutational analysis and for sequencing [22]. For a biologist, the obvious criteria for good choices were things like the GC content of the sequences since it is a good indicator of the melting temperature (at which the double helix dissociates) of short oligonucleotides. Arita and Kobayashi [22] introduced a template method to generate a set of sequences of length l such that any of its members have approximately $l/3$ mismatches (based on the GC content) with other sequences, their complements and the overlaps of their concatenations. Some approaches also tested combinatorial design, random generation, and genetic algorithms [18, 19]. A more refined approach considers the environment as an information channel that introduces undesirable changes (noise) to oligonucleotides (rather than bits) in transit to a double helix. DNA sequences diffuse in solution (the channel) "looking" for a (WC) complementary sequence (the destination) to hybridize to. If the probes used to encode data are not carefully selected, the equivalent of a transmission error occurs since the intended target may hybridize to the wrong probe (i.e., the hybridization affinity to the probe is not optimal). Thus, the [CWD] problem appears to be the equivalent of designing error-detecting/correcting codes in Shannon's information theory. Shannon's solution for error-detecting codes was to use the Hamming distance in Boolean hypercubes (described in Sect. 11.2) to quantify the error detecting and correcting capabilities by separating codewords actually used to encode single bits, so the noisy transmissions remain noncoding and can be detected and possibly corrected if they remain closer to the original codeword than to others. The obvious choice for [CWD] would thus appear to be the Hamming distance [23] between binary sequences. Since the Hamming distance between any two aligned sequences counts the number of nonmatching characters where the two do not match in a perfect alignment, the ordinary Hamming distance must be modified so that *matching* now refers to Watson–Crick complementary pairs, i.e., a's and t's (c's and g's) occurring in aligned sequences should be considered as matches. Some authors [24] used this notion of Hamming distance to obtain sets of "orthogonal" sequences solving the [CWD] problem experimentally and theoretically for molecular recognition using microarrays. Although this a step in the right direction, Hamming distance between two DNA strands appears to be too crude as an estimate of the likelihood of hybridization simply because it excludes the possibility of two strands hybridizing in shifted alignments [25], something much more likely to occur. To address this issue, an alternative was introduced in [17].

This h-distance turns out to be a reasonable choice for an approximation of the Gibbs energy because it satisfies metric properties that the Gibbs energy does not, and more importantly, because hybridization decisions made when the h-distance falls below an appropriate threshold τ agree with those made using the Gibbs energy given by the Nearest Neighbor Model (close to the actual decisions made by the real oligomers of length below 60 or so) about 80% of the time [6, 12].

Given an integer $m > 0$, the *h-measure* $h(x, y)$ between any two pmers x and y is the minimum total number of WC complementary mismatches between facing nucleotide pairs in an optimal alignment. Precisely, the computation of the h-measure proceeds as follows:

- Aligns x and y^r (y reversed) in $2n - 1$ alignments shifted by k characters (left shift if $k < 0$; right if $k > 0$), $-m < k < m$
- Counts the total number c_k of WC complementary mismatches between facing nucleotide pairs (single nucleotides are counted as mismatches)
- Returns for the h-measure the value

$$h(x, y) = \min_k \{c_k\} \, .$$

The *h-distance* (denoted just $|xy|$ henceforth) between two pmers x and y is defined as the minimum of the two h-measures $h(x, y)$ and $h(x, y')$ between x and y, where y' is the WC-complement of y [17].

The workflow for the computation of the h-distance is illustrated in Fig. 7.1.

DNA Metric Spaces
For every $m \geq 1$, the h-distance is a metric in $\mathbf{D}_m$, i.e., it is reflexive and symmetric and satisfies the triangle inequality. Furthermore, hybridization decisions between two mers x and y made when their h-distance falls below an appropriate threshold τ agree roughly about 80% of the time with one made using Gibbs' energy Nearest Neighbor Model with threshold $-6 \, kCal/mole$ [17, 25].

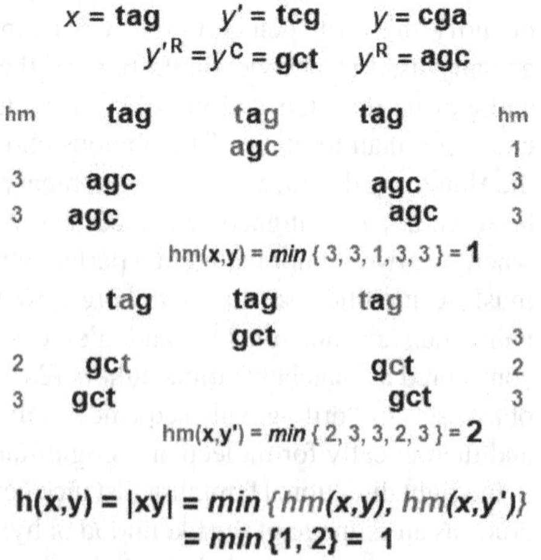

Fig. 7.1 Workflow for computing the h-distance $|xy|$ between any two pmers x and y by optimizing the frameshift for most WC-complementary (min of WC-nonmatching) pairs of facing nucleotides in the h-measure hm of the pairs x, y and x, y'. (Single nucleotides are considered mismatches.)

$$x = \text{tag} \quad y' = \text{tcg} \quad y = \text{cga}$$
$$y'^R = y^C = \text{gct} \quad y^R = \text{agc}$$

hm	tag	tag	tag	hm
		agc		1
3	agc		agc	3
3	agc		agc	3

$$hm(x,y) = min\{3, 3, 1, 3, 3\} = 1$$

	tag	tag	tag	
		gct		3
2	gct		gct	2
3	gct		gct	3

$$hm(x,y') = min\{2, 3, 3, 2, 3\} = 2$$

$$h(x,y) = |xy| = min\{hm(x,y), hm(x,y')\}$$
$$= min\{1, 2\} = 1$$

Table 7.2 Known isometries for DNA spaces $\mathbf{D}_m$ for $m \geq 1$, according to [26]. They are homomorphic substitutions named "bdef" to indicate that the characters in "acgt" in a pmer x/x' are mapped to those in "bdef," respectively. Any composition of two isometries is also an isometry

Name	Isometry ID	Mapping of acgt	Transformation
WC Complement/Identity	acgt	acgt	a↔t, c↔g Reverse+Complement
Polar ϕ_S^N	acgt	catg	a↔c, t↔g
Reverse ϕ^R	acgt	tgca	$\phi(x) = x^R$ (reverse pmer of x)
Polar + Reverse	acgt	gtac	Polar + Reverse
Polar2	acgt	gtac	a↔g, t↔c
g-c swap	acgt	agct	a↔a, c↔g, t↔t
a-t swap	acgt	tcga	a↔t, c↔c, g↔g

These metric properties of the h-distance can be used to solve the DNA CODEWORD DESIGN (**[CWD]**) problem *approximately*, so that data representations can be built and reasoned about as though they were physical objects like mass, centroids, and so on. Furthermore, it reveals some deeper structure of DNA hybridization landscapes, as follows:

An *isometry* ϕ of a DNA space $\mathbf{D}_m$ is an h-distance preserving transformation $\phi: \mathbf{D}_m \rightarrow \mathbf{D}_m$, i.e., for every pair of pmers $x, y \in \mathbf{D}_m$, $|\phi(x)\phi(y)| = |xy|$.

The deep structure of DNA can now be summarized more precisely by the following properties exhibited by every space of oligonucleotides of a fixed length, the $\mathbf{D}_m$ space (further details can be found in the original source [26]).

- *Every isometry ϕ of $\mathbf{D}_m$ must be injective and surjective. In particular, its inverse is also an isometry.*
- *The space $\mathbf{D}_m$ has at least 16 isometries.*

Table 7.2 shows the known isometries of DNA spaces. They are obtained by homomorphic (character by character) substitutions as shown. Thus, WC matchings are preserved and the h-measure remains unaffected upon substitutions. Therefore, they preserve the h-distance as well. (It is quite possible that other nonhomomorphic isometries are awaiting an explorer to discover them or show they are not possible).

These isometries afford an even more complete picture of the structure of the hybridization landscapes of oligomers of a given size (defined by the h-distance) through their images in DNA spaces. It is particularly interesting that this structure is something that humans are very familiar with in the planets Earth and Saturn because the structure of a DNA space $\mathbf{D}_m$ ($m \geq 2$) can be fully described by the following geometric concepts present in them:

- Two m-pmers x and y are an *antipodal* pair if and only if $|xy| = m$.
- The *north N* (*south S*) pole is the pmer a^m/t^m (c^m/g^m, respectively).
- The *northern ice cap* P^{m-1} is the set of pmers x satisfying $|xN| \leq 1$ and $|xS| = m$.
- The *equator* E_m is the set of all pmers x equidistant from the poles, i.e., satisfying $|xN| = |xS|$.

- The *northern hemisphere* (H^N) is the set of pmers x satisfying
  $|xN| < |xS|$.
- The images under the polar isometry ϕ of these objects are called the corresponding
  southern ice cap P_{m-1} and *southern hemisphere* (H_S).

Thus, the northern and southern equators are identical to the equator. The north
and south poles are an antipodal pair, and they partition the full DNA space $\mathbf{D}_m$.

The *parallels* of $\mathbf{D}_m$ are the following subsets of $\mathbf{D}_m$, for $1 \leq i \leq m$:
the ith *parallel* is the set P_i of all m-pmers x satisfying
$|xN| = R - i + \epsilon_N$ and $|xS| = R + i - \epsilon_S$,
where R is the maximum possible value of $|xN|$ and ϵ_N and ϵ_S are certain constants
depending on m and i.

E_m of $\mathbf{D}_m$ satisfies the following properties for every $m \geq 2$:

- *E_m is identical to the 0th parallel (P_0) with $\epsilon_N = \epsilon_S = 0$.*
- *E_m consists of (nearly) balanced m-pmers, i.e., the maximum number of occurrences of a's or t's is (nearly) identical to the maximum number of c's or g's.*
- *E_m is closed (hence invariant) under the polar and reversal isometries, i.e.,
  $\phi_S^N(E_m) \subseteq E_m$ and $\phi^R(E_m) \subseteq E_m$.*
- *$\mathbf{D}_m = H^N \cup E_m \cup H_S$.*

The *ellipse* with foci given by two m-pmers f_1 and f_2 and a constant $c \geq 0$ in a DNA space
$\mathbf{D}_m$ is the set of m-pmers x satisfying the condition $|xf_1| + |xf_2| = c$. The polar ellipse (as
shown in Fig. 7.2b) has foci at the poles N and S and the maximum possible value of c for
a nonempty ellipse.

*The polar ellipse is the nonempty ellipse with the largest c at least $2R$, where R is
the maximum possible distance of an m-pmer in the equator from the poles, and it
includes the whole equator E_m.*

The typical shape of the polar ellipse in DNA spaces is illustrated in Fig. 7.2b.

Let $k > 0$ be an integer and S be a set of pmers in $\mathbf{D}_m$ of size $|S|$ and w_z ($z \in S$) be a set
of real-valued weights for its elements. The (weighted) kth *error function* $SE_w^k D_n \to R$ is
defined as the average kth powers of the h-distances from z to a pmer in S, i.e., $SE_w^k(z) =
\frac{1}{|S|} \sum_{x \in S} w_z |zx|^k$. A pmer $\mathbf{c} \in D_m$ is a k-*centroid* of S (or simply *centroid* if $k = 2$) if and
only if it minimizes $SE_w^k(z)$ with all weights $w_k = 1$, i.e., $\mathbf{c} = \arg\min_z SE(z)$ across all z
in $\mathbf{D}_m$.

*An isometric image of a centroid in any DNA space $\mathbf{D}_m$ is also a centroid. There are
several centroids in every DNA space $\mathbf{D}_m$ for $m \geq 2$.*

These facts reveal why the shape of these spaces is very similar to that of planets
in our solar system (like Earth and Saturn), as shown in Figs. 7.2 and 7.3.

Such structural features can be computed by brute force for small values of
m ($m \leq 8$). But as the value of m increases, the size of $\mathbf{D}_m$ increases *exponentially*
(e.g., $\mathbf{D}_3$ has 32 pmers, $\mathbf{D}_5$ has 512 pmers, and $\mathbf{D}_8$ has 32,640 pmers), causing a
combinatorial explosion. Beyond $\mathbf{D}_8$, an exhaustive search of the space is practically

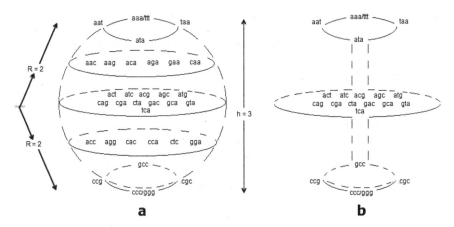

Fig. 7.2 Deep structure of DNA spaces $\mathbf{D}_m$ illustrated with $m = 3$. (a) The hybridization landscapes can be topologically sorted into positions on an equator, four parallels in the northern and southern hemispheres, the artic caps, and the north (N) and south (S) poles. (b) The polar ellipse (with foci at N and S) includes the full equator and the arctic caps (*in long dashes* around the poles). (This rendition is not isometric, although the relative separation in 3D Euclidean space between the location of the pmers is indicative of their actual h-distance, their likelihood of hybridization.)

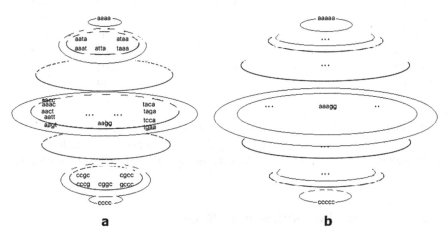

Fig. 7.3 Deep structure of DNA spaces $\mathbf{D}_m$ for (a) $m = 4$; and (b) $m \geq 5$. The hybridization landscapes can be topologically sorted into positions in structures reminiscent of the solar planet Saturn, with equatorial rings not only around the equator but alternate parallels as well, in both the north and southern hemispheres. This structure remains unaltered (except perhaps with more alternate rings around more parallels) for longer m-pmers. (This rendition is not isometric, although the relative separation in 3D Euclidean space between the location of pmers is indicative of their actual h-distance, their likelihood of hybridization. WC-palindromes have been excluded.)

impossible. Therefore, algorithms to generate all and only pmers satisfying the conditions in these properties would be useful for higher values of m. Some results for attributes for the smaller $\mathbf{D}_m$s ($m \leq 8$) are shown in Tables 7.3, 7.4, and 7.7.

Table 7.3 Some statistics about the parallels P_i in the DNA spaces $\mathbf{D}_m$ (given by their first lexicographic m-mers, e.g., ac stands for pmer ac/gt), according to [26]. The phenomenon of concentration of mass on the equator discussed in Sect. 3.3 is evident here too (percentages in parentheses)

m	$\lvert xN \rvert$	$\lvert xS \rvert$	$i / \epsilon_N, \epsilon_S$	$\lvert P_i \rvert$	P_i
3	0	3	3/1, 2	1	$P_3 = \{$ aaa $\}$
	1	3	2/1, 1	3	$P_2 = \{$ aat,ata,taa $\}$
	1	2	1/0, 1	6	$P_1 = \{$aac,aag,aca,aga, caa,gaa $\}$
	2	2	0/0, 0	12	$E_3 = \{$acg,act,agc,atc, atg,cag,cga,cta, gac,gca,gta,tca $\}$
4					$\lvert E_4 \rvert = 20 \ (\approx 16.7\%)$, as shown in Fig. 7.3
5					$\lvert E_5 \rvert = 120 \ (\approx 23.4\%)$
6					$\lvert E_6 \rvert = 580 \ (\approx 28.8\%)$
7					$\lvert E_7 \rvert = 1820 \ (\approx 22.2\%)$
8					$\lvert E_8 \rvert = 5832 \ (\approx 17.9\%)$

Table 7.4 Nxh bases for different DNA spaces [3, 20, 26]. Avg is the mean of the random variable N_τ counting the number of m-pmers x that hybridize to any probe in the basis, i.e., with $h(x, z) \leq \tau$. Its Entropy is $H(N_\tau)$. 4mP3 is the only basis with no hybridization uncertainty (every 4-pmer sticks to exactly one of the 3 probes for the given τ, i.e., N_τ is constant $N_\tau = 1$)

Nxh basis/chip	Length	Size	τ	Avg	Entropy
3mE4b	3	4	1.1	1.09	0.45
4mP3	4	3	2.1	1	0
6miC4Sa	6	4	4.1	1.03	0.26
7miC4Sa	7	4	4.1	1.02	0.17
7miC4Sb	7	4	4.1	1.02	0.15
8mP10	8	10	4.1	1.10	0.57

Further details about the deep structure of DNA spaces can be found in [26] and [6]. The structure can now be leveraged to perform exponential reductions in dimensionality of biotic and abiotic data, as shown in the remainder of this chapter.

7.2.2   Noncrosshybridizing (nxh) Bases

Microarrays have been the standard and popular tool to extract information from DNA sequences in biology. They consist of planar substrates such as glass, mica, plastic, or silicon, where DNA strands are affixed to allow specific bindings of bio-samples collected from an organism [27]. During the early 1990s, the first microarray experiments were performed using complementary DNA (cDNA) affixed to the microarrays. The length of a typical cDNA is 500–2500 base pairs, and they are widely used in gene expression assays [27]. Since 1990s, microarrays have been refined to capture and mine genomic and metabolomic data. The information gathered from this tool and data has wide applications in the fields of biology, medicine, health, and scientific research.

However, microarrays have serious disadvantages. First, analyses relying on their readouts give results that are hardly reproducible because of the high uncertainty in hybridization of targets to probes. The probes may not crosshybridize because they are affixed to the chip far apart, but the targets are floating in solution. No constraints are implemented in microarrays to minimize crosshybridization between targets. As a consequence, the results are not accurate and hence unreliable due to the lack of reproducibility, as argued in [28]. A second disadvantage of microarrays is that they may be too sparse and hence might miss target strands if they do not hybridize to any probe on the microarray and thus miss signals that could yield useful information.

Recent advances in next generation sequencing (NGS) have moved researchers away from microarrays directly to DNA fragments coding for proteins that can be used for processing and analysis instead. Currently, a number of NGS platforms using different sequencing technologies are available. These platforms perform sequencing of millions of small fragments of DNA in parallel in fairly short times. Some bioinformatic analyses join these fragments by mapping the individual reads to a reference genome [29]. However, analyzing the sequences generated using these platforms is a big challenge. Through the use of deep learning models on these sequences directly to extract useful predictors automatically, the disadvantages of microarray analyses could be reduced (no risk of unwanted hybridization as the phenomenon is not considered at all). However, the performance of such networks is highly dependent on the quality and/or relevance of the data as well as the size of the data. In particular, such models can pretty much memorize data when it is limited. In fact, there is an ongoing debate between two extreme approaches in the field of machine learning (i.e., feed raw data to a model without any processing to avoid bias vs manual selection of features that might be important), and the researchers have concluded that there should be a trade-off [30]. Furthermore, the results obtained using these methods are not explainable since it is not clear how the model is making decisions (e.g., non/cancerous) that would allow a human to rationalize and accept or reject the decisions.

Using the deep structure of DNA spaces, an entirely different approach to dimensionality reduction emerges. A selected set of nxh pmers (nxh bases) could be used to reduce the dimension of DNA sequences and thus extract more relevant information about the sequences based on the knowledge of Gibbs energy landscapes. These pmers will be referred to as *probes* (contrary to the standard use in biology where they are referred to as targets). It has been shown [6, 31] that the **[CWD]** problem can be reduced to a well-known and well-researched problem in geometry, a *sphere-packing* problem. With some additional work, solutions to the **[CWD]** problem can be had for values of $m \leq 8$. Table 7.4 shows some of the nxh bases used below for DR and their quality.

There are several advantages of using these strategically selected sets of probes (called *noncrosshybridizing (nxh) bases*) with microarrays and NGS. First, nxh bases can be used to transform arbitrary DNA sequence into numerical feature vectors. These vectors could be used to train conventional statistical and machine learning models like regression models, support vector machines, random forests, decision trees, and multilayer perceptrons (defined in Sect. 2.2). The drawbacks of deep networks requiring abundant data for effective learning can be avoided with the use of these models based on nxh features, as shown below. Furthermore, the results obtained using these bases can be rationalized because they reflect deep knowledge of the Gibbs energy landscapes of DNA hybridization critical to living organisms. Hence, the results will be more explainable.

7.3 Reduction by Genomic Signatures

This section introduces a second kind of DR based on the pointwise hybridization pattern exhibited by a dataset encoded as DNA sequences to a common judiciously selected set of DNA oligonucleotides (an nxh DNA basis or DNA chip) of the same length blanketing the entire DNA space. A classic problem of species identification or biotaxonomic classification **[BioTC]** and malware classification **[MalC]** (both defined in Sects. 1.2 and 11.4) are used as running examples and are further discussed next.

Example 7.3 Biotaxonomic classification is a classical problem in biology. The significance of and challenges surrounding this problem were briefly discussed earlier in this chapter. Although these biological challenges fall outside of the scope of this book, the usage of genomic sequences in statistical or machine learning models serves as a suitable application of dimensionality reduction on genomic data. One way to extract information from DNA sequences of living organisms would be to do a homomorphic substitution of DNA characters by numbers (as described above in Sect. 7.1) and then use convolutional networks to make some predictions about their taxonomic groups. The disadvantage of using this approach is that DNA sequences can be very long. For example, on average, the DNA sequence of a gene like *cytochrome c oxidase subunit I* of organisms like blackfly contains about 600–700 DNA oligonucleotides [32]. This number can increase drastically if the sequences involved include the entire mitochondria or even, ideally, the whole genome. So, using even 600 features requires advanced models like convolutional networks demanding a lot of computational resources. On the other hand, identifying some short DNA oligos very important in the context of the geometric properties of DNA spaces (as discussed in the previous section) and computing a pattern of hybridizing affinities of DNA sequences in a sample with these short DNA oligos might help to reduce the dimensionality of these longer DNA sequences. The corresponding features could then be fed to standard statistical and/or machine learning models.   □

Example 7.4 Molecular dimensionality reduction can be applied to abiotic data (having nothing to do with living organisms, e.g., malwares) as well. Naturally, that requires the data to be coded into DNA first, as discussed above in Sect. 7.1.   □

The conceptual framework to perform such dimensionality reduction is based on the deep structure of DNA described in Sect. 7.3.2 and is described in this section.

7.3.1   Background

As a blueprint of life, DNA encodes critical information quinta-essential for the development and survival of a living organism (e.g., protein synthesis) [1]. Its properties allow the cell to store and process all kinds of information that determine not only morphology and phenotype but also the metabolic behavior of an organism. Prime examples are genes like BRCA1 and BRCA2, which play an important role in breast and ovarian cancers caused in human females 60% of the time. Early genetic studies showed that DNA is responsible for preservation of most phenotypes from parents (specifying their structure and function) into their offspring; further studies have probed into the role of epigenetics in disease [33], e.g., Jin and Liu [34] point out that DNA methylation is one of the earliest such modifications in humans. This type of alteration is linked with many cancer types (such as colon, breast, liver, bladder, Wilms, ovarian, esophageal, prostate, and bone), autoimmune diseases, metabolic disorders, neurological disorders, and so forth. For example, clinical sequencing has demonstrated its effectiveness in serving as an alternative to identify diseases caused by genetic factors. Furthermore, there is evidence now that a genome may encode even information about environmental habitat where an organism grew and lived [35].

In order to enable extraction of this information from DNA sequences, they must be pre-processed. Microarrays have been the standard and popular tool to do so in biology [27]. However, as pointed out above, the unreliability and irreproducibility of the results (due to the noise present in microarray signals) led to current advances in sequencing technologies (e.g., next generation sequencing, NGS). Enormous progress in NGS in the last two decades has substantially lowered the cost and time required for genome sequencing for any organism. But the major challenge remains how to extract and interpret the huge amount of information about the organism contained in these sequences to turn this data into useful and explainable outcomes [36]. As it turns out, structural properties of DNA enable an even deeper and more informative analysis. Judicious selections among the centroids of the parallels in DNA spaces (as described in Sect. 7.2) afford designs of nxh bases and help address these drawbacks of microarrays, as shown in Table 7.4.

7.3.2   Genomic Signatures

An nxh basis can be used to transform a genomic sequence into a numerical vector that could be used to train machine learning models to solve challenging problems.

Given an nxh basis B of probe length $m > 1$ with k probes, a hybridization threshold $\tau > 0$, and a DNA sequence x of any length $x \geq m$, the mD *genomic signature* of x on B at h-threshold τ is obtained by an algorithm that

- Shreds x to nonoverlapping fragments of size m (ignoring any shorter leftover shreds, if any)

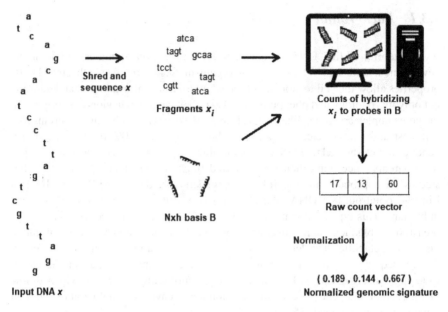

Fig. 7.4 Workflow to compute the genomic signature of an arbitrary DNA sequence x (perhaps from a living organism) using an nxh basis B

- Computes the total number of shreds in x that hybridize with each probe $z_i \in B$ at threshold τ
- Normalizes the mD vector obtained using the partition function (i.e., dividing all components by the total number of shreds given by the nearest integer $\lfloor |x|/m \rfloor$ at or below $|x|/m$)

This workflow to compute such a signature is illustrated in Fig. 7.4. Perl and Python scripts implementing this algorithm were used to shred an arbitrary DNA sequence into fragments of probe length. For each probe in a given basis, the total number of shreds in the sequence hybridizing with it given a stringency condition τ is counted. The hybridization criterion is that two oligos of the same length hybridize with each other if and only if their Gibbs energy, as approximated by their h-distance, is less than τ; this decision agrees with the decision based on Gibb's energy Nearest Neighbor Model about 80% of the time [6], as pointed out in Sect. 7.2. The normalization of this vector is the genomic signature of the DNA sequence. The signatures in a taxon of organisms capture very deep and useful information about them (e.g., information to predict phenotypic features and environmental features in the natural habitat of blackfly and *Arabidopsis thaliana* [3, 4]) despite the enormous dimensionality reduction (from thousands and possibly millions or billions of nucleotides to a vector of as many dimensions as the number of the few probes in the nxh basis, as illustrated in Table 7.4).

The quality of the information content of these signatures can be traced back to the quality of the nxh bases used and the information content preserved in the dimensionality reduction. The quality of nxh bases can be evaluated by using two quantitative metrics, i.e., Shannon entropy and the average number of pmers a probe is likely to hybridize. To compute the entropy of a basis, a random experiment can be performed (as suggested in [35]). The experiment consists of randomly choosing a pmer to check for hybridization with any probe. One good choice of random variable observes the total number of probes in a basis that the random pmer hybridizes with. If the variable is constant and its value is certain to occur, then there is no uncertainty and $H(X) = 0$s, as is the case with the basis 4mP3 for $\mathbf{D}_3$. On the other hand, if the values are about equally likely to occur, the variable has maximum uncertainty and $H(X) \geq 1$ (as is the case for a random selection of elements in a crosshybridizing set). The entropy of the bases used for solutions and further assessment below is shown in Table 7.4.

The quality of the information extracted by an nxh basis can be illustrated by the fact that, surprisingly, genomic signatures actually capture even nongenetic information encoded in DNA, for example, about the location and average temperature of the environment where an organism lived. The genomic signatures of each data point in the datasets for the [MalC] and [BioTC] problems can be computed following the algorithm described above. An additional process of encoding malwares into DNA sequences is required for the [MalC] problem. Since the genomic signatures are so short, it is not surprising that the solutions obtained using these signatures can be processed much faster than, say, CNNs networks would. The only time-consuming step in the entire process is cleansing DNA sequences (i.e., removing non-DNA characters in DNA sequences files in the dataset for BioTC problem; no such cleansing is needed for the [MalC] problem). Even for the [BioTC] problem, the process was completed under an hour and is a one-time event for a particular data point). Then, the entire process of computing genomic signatures and training models to solve the problems can be performed in the order of minutes for all datasets in the assessment next.

Therefore, the critical question is how good these solutions to the chosen problems [MalC] and BiotC] are. The scores are given in Tables 7.5 and 7.6. In order to assess them, these models were compared against the state-of-the-art solutions for these problems, shown below in Table 7.9. For the problem [BiotC] of biotaxonomic classification, several solution models were proposed by the authors in [40], namely Naive Bayes, Random Forests, k-Nearest Neighbors, and DeepBarCoding (based on Deep Networks). The best score was obtained by DeepBarCoding (F1-Score = 0.9763). The average across all models is 92.8 and the standard deviation is 0.0475. All solution models with the scores above 92.88% are thus of acceptable quality, namely Multiple Regression Models and Random Forests trained using genomic signatures on basis 3mE4b. A similar comparison for the [MalC] problem is shown together with another dimensionality reduction technique in Sect. 7.4 next.

Table 7.5 Performance of machine learning models when trained using genomic signatures to solve the **[BioTC]** problem

Nxh basis	Method	Precision (%)	Recall (%)	F1-Score (%)
3mE4b	k-Means	93.80	95.00	92.80
	LDA	88.31		
	MLR	93.26		
	SVM	67.05		
	GNB	84.84		
	RF	95.89		
	kNN	91.82		
	NN	66.90		
4mP3	k-Means	91.30	92.20	89.30
8mP10	LDA	43.16		
	MLR	49.14		
	SVM	45.44		
	GNB	41.61		
	RF	49.49		
	kNN	44.90		
	NN	42.79		

Table 7.6 Performance of machine learning models when trained using genomic signatures to solve the **[MalC]** problem and comparison to the state-of-the-art solutions

Nxh basis	Method	Precision (%)	Recall (%)	F1-Score (%)
	Kaggle Winner [37]	99.63	99.07	99.35
	SNNMAC [38]	99.21	99.18	99.19
	MalNet [39]	99.14	97.96	99.55
	RF [38]	84.46	82.34	83.38
	LR [38]	71.42	67.38	69.34
	SVMs [38]	54.84	28.75	37.72
3mE4b	RF	99.33	99.05	99.13
	kNN	94.42	95.19	94.11
8mP10	RF	73.22	74.79	73.34
	kNN	71.08	70.85	69.86

7.4 Reduction by Pmeric Signatures

This section describes another reduction of DNA data based on the barycentric coordinates of the Euclidean representation of the h-distance centroids of a DNA space $\mathbf{D}_m$. This reduction captures their hybridization affinity in terms of their h-distances to some special pmers that have the capability of representing arbitrarily long DNA sequences.

Table 7.7 Centroids for the DNA spaces D_m of all pmers of length m (as given by their first lexicographical m-mers, e.g., ac stands for pmer ac/gt), as described in [26]

m	h-centroids
3	aca, aga, cac, ctc
4	acca, agga, caac, cttc
5	accat, aggat, caacg, cgaag, gaagc, gcaac, tacca, tagga
6	acaagc, agaacg, atccac, atggag, caccta, cgaaca, gaggta,gcaaga
7	actccat, agtggat, cagaacg, cgaacag, gacaagc, gcaagac,tacctca, taggtga
8	actatccg, agtatggc, attcgctg, attgcgtc, cggtatga, ctgcgtta, gcctatca, gtcgctta

Example 7.5 The problems of biotaxonomic classification **[BioTC]** and malware classification **[MalC]** (along with the corresponding datasets described in Sect. 11.5) can again be used to demonstrate the effectiveness of this dimensionality reduction technique.   □

The conceptual framework necessary to understand this technique is based on the concept of centroid in a DNA space defined in Sect. 7.2. These centroids are shown in Table 7.7 for small $m \leq 8$. It is motivated by the question, what is the center of the space D_m? Since DNA spaces look spherical, the question makes some sense, e.g., what point is closest to all other pmers on the average (within the smallest radius)?

These centroids serve as an alternative way to transform DNA sequences into lower dimensional feature vectors, the so-called *pmeric signatures*. The key idea is that a point in a metric space may be determined by its distances from certain key points, as is the case of Euclidean spaces in Cartesian coordinates (signed distances from two fixed lines, the x- and y-axis) or polar coordinates (a distance from the origin and a polar angle with the x-axis). There is no obvious concept of a line or an angle in D_m, but the centroids act as natural origins. (Unlike on Earth where all objects are attracted toward a unique center of mass due to gravitational forces, there is no unique h-centroid in DNA spaces due to the symmetries (isometries) of D_m). Furthermore, more than one pmer might share the same coordinates, i.e., pmeric coordinates do not necessarily determine a point uniquely. However, the number of appearances of these pmers in genomic sequences of different strings is likely to be different if masses are placed at a pmer of size equal to the ratio of the total number of times the pmer occurs in x to the total number of m-pmer shreds in x. Thus, distinguishing several organisms/abiotic datapoints (e.g., images) based on these vectors is still possible. These vectors will be used as the so-called pmeric signatures for the respective strings, defined precisely as follows:

Let k be the number of centroids in D_m and x be a DNA string of size $n > 1$. The *pmeric signature* of x is the mD vector obtained by:

- Shredding x into nonoverlapping fragments of size m (ignoring any shorter leftover fragments, if any)

- Computing $y_{ij} = w_j |z_i x_j|$, for each centroid $z_i \in \mathbf{D}_m$ and for each unique shred x_j, where w_j is the fraction of the number of occurrences of x_j to the total number of shreds in x
- Taking for ith component of the pmeric signature of x the weighted average (given by the sum of the y_{ij} s) across all shreds x_j

A DNA sequence from a living organism can be sequenced to produce the corresponding DNA string. A Python script can be written to shred DNA strings into pmers of uniform length m. In particular, each pmer in $\mathbf{D}_m$ can be viewed as a point with certain weights given by its h-distances from the h-centroids in $\mathbf{D}_m$. Although the pmeric signatures hardly determine the pmer uniquely because the pmer is too short, longer sequences provide weights to distinguish them.

Example 7.6 The pmeric signatures of malwares can be computed using their DNA encodings (as described in Sect. 7.1). These pmeric signatures can then be used as feature vectors for training several machine learning models for a given problem. The performance scores for **[MalC]** and **[BioTC]** are shown in Table 7.8.   □

Again, an important question arises about the quality of these models. To assess it, the performance scores as reported below in Table 7.8 can be compared to some state-of-the-art solution methods. From Tables 7.6 and 7.8, it is clear that the performance scores of the dimensionality reduction method based on genomic signatures using nxh basis 3mE4b are nearly equal to the ones of winner of the challenge (as reported on *kaggle.com*). All performance scores (except for the DR methods based on genomic signature using 8mP10) are greater than the difference between the corresponding performance score of the winner and the standard deviation of the scores from the literature (i.e., the threshold for $precision = 83.93$, $recall = 76.42$, and $F1\text{-}score = 79.27$). Therefore, these models are of better than acceptable quality. In fact, given that only 4D vectors are being used with 3mE4b after the dimensionality reduction, it is very surprising that these solutions to malware classification are very competitive with the performance of the winner of the Microsoft challenge (which involved over 300 selected features out of 1804) (Table 7.9).

Table 7.8 Performance of pmeric signature features on machine learning models trained for the malware classification problem **[MalC]**

Size m	Method	Precision (%)	Recall (%)	F1-Score (%)
8pmc	RF	95.83	95.75	96.62
	kNN	95.41	95.24	96.38

Table 7.9 The performance scores of the *state-of-the-art* solutions for Biotaxonomical Classification and Malware Classification

Nxh basis	Method	Precision (%)	Recall (%)	F1-Score (%)
Related work for [MalC]				
	Kaggle Winner [37]	99.63	99.07	99.35
	SNNMAC [38]	99.21	99.18	99.19
	MalNet [39]	99.14	97.96	98.55
	In [41]	92.13	90.64	91.38
Machine learning models [38]				
	Random Forest	84.46	82.34	83.38
	Xgboost	85.13	72.02	78.02
	Naive Bayes	70.21	70.06	70.13
	Logistic Regression	71.42	67.38	69.34
	Support Vector Machine	54.84	28.75	37.72
Related work for [BioTC] [40]				
	SVM			0.9688
Machine learning models				
	Naive Bayes	–	–	0.8558
	RF	–	–	0.9690
	kNN	–	–	0.9036
	NN	–	–	0.9345
	DeepBarCoding	–	–	0.9763

References

1. Watson J. D., & Crick, F. H. C. (1953). Molecular structure of nucleic acids: a structure for deoxyribose nucleic acid. *Nature, 171*(4356), 737–738.
2. Mainali, S., Garzon, M., & Colorado, F. A. (2020). New genomic information systems (GenISs): Species delimitation and identification. In *International Work-Conference on Bioinformatics and Biomedical Engineering* (pp. 163–174). Springer.
3. Mainali, S., Colorado F. A., & Garzon M. H. (2021). Foretelling the phenotype of a genomic sequence. *IEEE/ACM Transactions on Computational Biology and Bioinformatics, 18*(2), 777–783.
4. Mainali, S., Garzon, M., & Colorado F. A. (2020). Profiling environmental conditions from DNA. In *International Work-Conference on Bioinformatics and Biomedical Engineering* (pp. 647–658). Springer.
5. Adleman Leonard, M. (1994). Molecular computation of solutions to combinatorial problems. *Science, 266*(5187), 1021–1024.
6. Garzon, M. H., & Bobba, K. C. (2012). A geometric approach to Gibbs energy landscapes and optimal DNA codeword design. In *International Workshop on DNA-Based Computers* (pp. 73–85). Springer.
7. Seeman N. C. (2003). DNA in a material world. *Nature, 421*(6921), 427–431.
8. Linnaeus, C. (1758). *System naturae* (Vol. 1). Stockholm Laurentii Salvii.

9. Kumar, S., Stecher, G., Suleski, M., & Hedges S. B. (2017). TimeTree: a resource for timelines, timetrees, and divergence times. *Molecular Biology and Evolution, 34*(7), 1812–1819.
10. Wake, M. H. (2008). Integrative biology: Science for the 21st century. *BioScience, 58*(4), 349–353.
11. Mizrachi, I. (2007). GenBank: the nucleotide sequence database. *The NCBI handbook [Internet], updated*, 22.
12. Garzon, M. H., Bobba, K., Neel, A., & Phan, V. (2010). DNA-based indexing. *International Journal of Nanotechnology and Molecular Computation (IJNMC), 2*(3), 25–45.
13. Neel, A. J., & Garzon, M. H. (2008). DNA-based memories: a survey. In *New developments in formal languages and applications* (pp. 259–275). Springer.
14. Neel, A., & Garzon, M. H. (2012). Semantic methods for textual entailment. In *Applied natural language processing: Identification, investigation and resolution* (pp. 479–494). IGI Global.
15. Goldman, N., Bertone, P., Chen, S., Dessimoz, C., LeProust, E. M., Sipos, B., & Birney, E. (2013). Towards practical, high-capacity, low-maintenance information storage in synthesized DNA. *Nature, 494*(7435), 77–80.
16. Winfree, E., Liu, F., Wenzler L. A., & Seeman N. C. (1998). Design and self-assembly of two-dimensional DNA crystals. *Nature, 394*(6693), 539–544.
17. Garzon, M., Neathery, P., Deaton, R., Murphy, R. C., Franceschetti, D. R., & Stevens Jr., S. E. (1997). A new metric for DNA computing. In *Proceedings of the 2nd Genetic Programming Conference* (pp. 472–478). Morgan Kaufman.
18. Frutos, A. G., Condon, A., & Corn, R. (1997). Demonstration of a word design strategy for DNA computing on surface. *Nucleic Acids Research, 25*, 4748–4757.
19. Deaton, R., Garzon, M., Murphy, R. C., Rose, J. A., Franceschetti, D., & Stevens Jr., S. E. (1998). The reliability and efficiency of a DNA computation. *Physical Review Letters, 80*, 417.
20. Garzon, M. H., & Mainali, S. (2017). Towards reliable microarray analysis and design. In *9th International Conference on Bioinformatics and Computational Biology (ISCA)* (6 pp.).
21. Wetmur, J. G. (1997). Physical chemistry of nucleic acid hybridization. In *DIMACS series in discrete mathematics* (vol. 48, pp. 1–23).
22. Arita, M., & Kobayashi, S. (2002). DNA sequence design using templates. *New Generation Computing, 20*(3), 263.
23. Roman, J. (1995). *The theory of error-correcting codes* (1st ed.). Springer-Verlag.
24. Mohammadi-Kambs, M., Hölz, K., & Somoza, M. M. (2017). Hamming distance as a concept in DNA molecular recognition. *ACS Omega, 2*, 1302–1308.
25. Phan, V., & Garzon Max, H. (2009). On codeword design in metric DNA spaces. *Natural Computing, 8*(3), 571.
26. Garzon, M. H., & Mainali, S. (2021). Deep structure of DNA for genomic analysis. *Human Molecular Genetics, 31*(4), 576–586. https://doi.org/10.1093/hmg/ddab272
27. Schena, M. (2003). *Microarray analysis*. Wiley-Liss.
28. Garzon, M. H., & Mainali, S. (2017). Towards a universal genomic positioning system: phylogenetics and species identification. In *International Conference on Bioinformatics and Biomedical Engineering* (pp. 469–479). Springer.
29. Behjati, S., & Tarpey P. S. (2013). What is next generation sequencing? *Archives of Disease in Childhood-Education and Practice, 98*(6), 236–238.
30. Marcus, G. (2018). *Innateness, AlphaZero, and Artificial Intelligence*. Preprint. arXiv:1801.05667.
31. Garzon, M. H. (2014). DNA codeword design: Theory and applications. *Parallel Processing Letters, 24*(02), 1–21.

32. Colorado-Garzón, F. A., Adler, P. H., García, L. F., Muñoz de Hoyos, P., Bueno, M. L., & Matta, N. E. (2017). Estimating diversity of black flies in the Simulium ignescens and Simulium tunja complexes in Colombia: chromosomal rearrangements as the core of integrative taxonomy. *Journal of Heredity, 108*(1), 12–24.

33. Cook-Deegan, R., DeRienzo, C., Carbone, J., Chandrasekharan, S., Heaney, C., & Conover, C. (2010). Impact of gene patents and licensing practices on access to genetic testing for inherited susceptibility to cancer: comparing breast and ovarian cancers with colon cancers. *Genetics in Medicine, 12*(1), S15–S38.

34. Jin, Z., & Liu, Y. (2018). DNA methylation in human diseases. *Genes & Diseases, 5*(1), 1–8.

35. Mainali, S., Garzon, M., Venugopal, D., Jana, K., Yang, C. C., Kumar, N., Bowman, D., & Deng, L. Y. (2021). An information-theoretic approach to dimensionality reduction in data science. *International Journal of Data Science and Analytics, 12*, 1–19.

36. Sun, H., & Yu, G. (2019). New insights into the pathogenicity of non-synonymous variants through multi-level analysis. *Scientific Reports, 9*(1), 1–11.

37. Wang, X., Liu, J., & Chen, X. (2015). Microsoft malware classification challenge (big 2015) first place team: say no to overfitting. In *No. Big*.

38. Yang, P., Zhou, H., Zhu, Y., Liu, L., & Zhang, L. (2020). Malware classification based on shallow neural network. *Future Internet, 12*(12), 219.

39. Yan, J., Qi, Y., & Rao, Q. (2018). Detecting malware with an ensemble method based on deep neural network. *Security and Communication Networks,* **2018**, Article ID 7247095. https://doi.org/doi.org/10.1155/2018/7247095

40. Yang, C. H., Wu, K. C., Chuang, L. Y., & Chang, H. W. (2021). DeepBarCoding: Deep learning for species classification using DNA barcoding. *IEEE/ACM Transactions on Computational Biology and Bioinformatics*.

41. Zhang, H., Xiao, X., Mercaldo, F., Ni, S., Martinelli, F., & Sangaiah A. K. (2019). Classification of ransomware families with machine learning based on n-gram of opcodes. *Future Generation Computer Systems, 90*, 211–221.

Chapter 8
Statistical Learning Approaches

Ching-Chi Yang (iD) **and Lih-Yuan Deng** (iD)

Abstract Instead of retaining certain properties when selecting or extracting features, other methods aim to remove irrelevant and/or redundant features in the data using primarily statistical criteria. Features are now selected or extracted that have the highest impact on the prediction of the response/target variable based on various statistical solution methods. This chapter describes methods using linear regression and regularization that afford solutions to dimensionality reduction and solutions to problems that are explainable to humans.

In Chap. 6, information-theoretic methods were used to reduce dimensionality in a dataset aiming to maximize information in the selected features by pre-processing the dataset. This chapter uses a similar idea but using various statistical methods to solve data science problems (such as regression and statistical learning) for dimensionality reduction.

8.1 Reduction by Multiple Regression

One can infer relationships between features by building a linear regression model with a straight forward interpretation. If a feature X_i in the dataset **X** can be inferred from other features, then X_i is probably redundant and can be discarded if the other features are included, just as it was done with methods based on entropy in Chap. 6.

Example 8.1 (Multiple Linear Regression for the Body Fat Dataset) Table 8.1 shows the coefficients of the input variables under multiple linear regression for the body fat. One can then interpret the coefficients to make inferences. For example, the coefficient 0.06 of the feature Age implies that a person one year older can be expected to have 0.06 units more of body fat on average, assuming the values of the

C.-C. Yang (✉) · L.-Y. Deng
Mathematical Sciences, The University of Memphis, Memphis, TN, USA
e-mail: cyang3@memphis.edu; lihdeng@memphis.edu

© The Author(s), under exclusive license to Springer Nature Switzerland AG 2022   169
M. Garzon et al. (eds.), *Dimensionality Reduction in Data Science*,
https://doi.org/10.1007/978-3-031-05371-9_8

Table 8.1 Coefficients for a multiple linear regression for the target feature in the Body Fat dataset with the remaining features as predictors

Variable	Coefficient	Variable	Coefficient
(Intercept)	−18.07	Hip	−0.21
Age	0.06	Thigh	0.23
Weight	−0.09	Knee	0.03
Height	−0.06	Ankle	0.17
BMI	0.07	Biceps	0.18
Neck	−0.48	Forearm	0.45
Chest	−0.03	Wrist	−1.62
Abdomen	0.95		

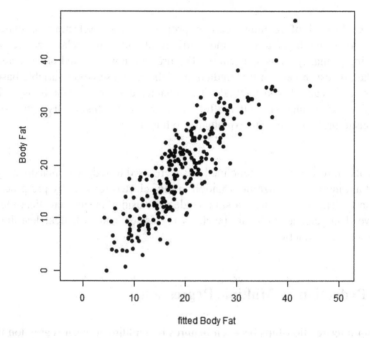

Fig. 8.1 The predicted values (*x*-axis) for the body fat dataset are fairly consistent (near the diagonal) with the observed values (*y*-axis)

remaining variables remain fixed. Figure 8.1 shows the predicted values with the observed values for the dataset. Since the scatter plot clusters about the diagonal, the fitted linear model performs reasonably well. □

In general, the dataset contains n observations $(\mathbf{x}_1, y_1), \ldots, (\mathbf{x}_n, y_n)$, where $\mathbf{x}_i = (x_{i1}, \ldots, x_{ip})$ is a covariate vector (predictors/independent variables) for the i th observation and y_i is the corresponding response (as described in Chap. 1.) Without loss of generality, one can assume that the response variable is a scalar type and is not included in $\mathbf{X}$. The dataset can then be represented as a matrix $\mathbf{X}$ of size $n \times p$

$$\mathbf{X} = \begin{bmatrix} x_{11} \; \cdots \; x_{1p} \\ x_{21} \; \cdots \; x_{2p} \\ . \\ . \\ x_{n1} \; \cdots \; x_{np} \end{bmatrix} = [X_1, X_2, \cdots, X_p], \quad \mathbf{Y} = \begin{bmatrix} y_1 \\ y_2 \\ . \\ . \\ y_n \end{bmatrix} \tag{8.1}$$

where the covariate matrix can be viewed as n row vectors $(\mathbf{x}_1, \mathbf{x}_2, \cdots, \mathbf{x}_n)$ of dimension p (the data points) or p column vectors of dimension n $(X_1, X_2, \cdots, X_p)$ (the features) and $\mathbf{Y}$ is the response column vector of dimension n, the transpose $(y_1, y_2, \ldots, y_n)'$. The matrix form of the multiple linear model is

$$\mathbf{Y} = \mathbf{X}\beta + \varepsilon, \tag{8.2}$$

where $\varepsilon \sim N(\mathbf{0}, \sigma^2\mathbf{I})$ and $\mathbf{I}$ denotes the identity matrix. This matrix form can also be written as a linear model for the i th observation as

$$y_i = \sum_{j=0}^{p} \beta_j x_{ij} + \varepsilon_i. \tag{8.3}$$

for $i = 1, 2, \cdots, n$, where the ε_is are random error variables with i.i.d. $N(0, \sigma^2)$.

The unknown parameter β can estimated using the Ordinary Least Squares (OLS) method, i.e., so that the sum of squares of residuals is minimized using the loss function

$$L(\beta) = \|(\mathbf{Y} - \mathbf{X}\beta)\|_2^2 = (\mathbf{Y} - \mathbf{X}\beta)'(\mathbf{Y} - \mathbf{X}\beta) .$$

The best solution is the least square estimator (LSE) for β given by

$$\hat{\beta} = (\mathbf{X}'\mathbf{X})^{-1}\mathbf{X}'\mathbf{Y}.$$

In many situations where many features can be used to predict a response variable (e.g. with body fat), many of the features may have no effect on the response. Including these variables in a subsequent model causes the analysis to be overly complex without a return in useful information. One simple (but not very useful) way to reduce the number of input variables is to perform a simple t-test on all p variables individually. Other more effective procedures to reduce the number of effective input variables are described next.

8.2   Reduction by Ridge Regression

Before LASSO regression (described in Sect. 8.3) became popular, the Ridge Regression method was one of the most popular methods. It is especially useful when the number of input variables (p) is large and the whole data matrix may is of nonfull rank, hence exhibits high multicollinearity or correlations among features. Its main goal is to lower the size of the coefficients to avoid over-fitting, while keeping them away from 0. Consequently, the final model is less interpretable because it still has a large number of variables.

When there are more features than data points ($p > n$), the OLS procedure for Multiple Linear Regression (MLR) will fail because the matrix ($\mathbf{X'X}$) is singular (no inverse exists.) Even if nonsingular, if $p < n$ and the independent variables are highly correlated, the matrix will be too close to singular to make MLR haphazardous. As an alternative solution to the issue of least square estimators when the data exhibits some highly correlated input variables or predictors, Hoerl and Kennard [1] offered *Ridge Regression*. The most common form uses a ℓ_2-penalty of the form

$$L(\beta) = \|(\mathbf{Y} - \mathbf{X}\beta)\|_2^2 + \lambda\|\beta\|_2^2.$$

The best solution is the ridge estimator of β given by

$$\hat{\beta} = (\mathbf{X'X} + \lambda\mathbf{I})^{-1}\mathbf{X'Y}.$$

Ridge regression uses the ℓ_2-norm to penalize a larger number of input variables. LASSO regression is another type of penalized regression that uses the ℓ_1-norm. The main advantage of LASSO Regression is that it can provide a more precise estimate for the ridge parameters, as its variance and mean square estimator are often smaller than the least square estimators of Sect. 8.1. Compared with LASSO, ridge regression usually does not provide a significant dimensionality reduction by reducing the variable's coefficient to 0s. By conservatively suggesting a bigger set than LASSO regression does, researchers can be more confident to remove the unselected variables. Because of the trade-off between LASSO and Ridge, people usually consider a mixture penalty of LASSO and Ridge penalties.

To summarize, LASSO works better when one has more features and needs to make a simpler and more interpretable model. However, if the features in the data have high correlation, LASSO may not work well and Ridge Regression could be a better choice. It is also possible to combine the two in the Elastic Net method described below in Sect. 8.5.

8.3   Reduction by Lasso Regression

LASSO regression is another type of penalized regression. It uses the ℓ_1-norm (instead of the ℓ_2-norm in Ridge regression) to penalize using a larger number of predictor variables .

Example 8.2 (LASSO Regression with the Body Fat Dataset) Figure 8.2 shows the coefficients of the predictor variables for various values of the penalty (λ). The top of the figure shows the number of variables recommended for different values of λ. If the penalty term is larger, fewer variables have nonzero coefficients. By comparison with the example in Sect. 8.3, Ridge regression still recommends the same number of 14 variables for a different penalty λ. □

Depending on the loss function f (defined in Chap. 2), the ℓ_1 regularization can be generally written as

$$L(\beta) = \|(\mathbf{Y} - \mathbf{X}\beta)\|_2^2 + \lambda \sum_{i=1}^{p} |\beta_i|$$

where λ is a regularization weight. For a given λ, minimizing the penalized loss function reduces some variables' coefficients to 0. When λ becomes sufficiently large, LASSO gives a null model because all coefficient estimates become 0.

Advantages of using a LASSO reduction include

1. feature selection;
2. it is much easier to interpret and produces simpler models;

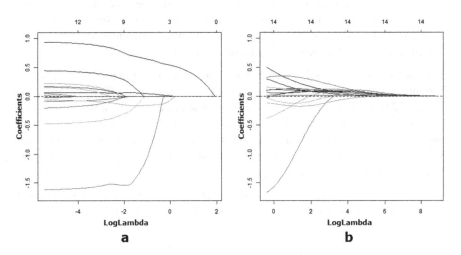

Fig. 8.2 Coefficients of the predictor variables for various values of the penalty λ obtained using (**a**) multiple LASSO regression; and (**b**) multiple Ridge regression

3. performs better in a setting where a relatively small number of predictors have large coefficients;
4. the remaining predictors have coefficients that are very small or equal 0;
5. LASSO tends to outperform Ridge regression in terms of bias, variance, and MSE of the solutions.

In practice, a LASSO penalty provides many advantages and can be used in many other loss functions. However, there are also disadvantages of using LASSO in a number of research areas, including

1. if $n < p$ (high dimensional case), LASSO can select at most n features because of the nature of convex optimization LASSO relies on;
2. for correlated features (as is the case in most of the real word datasets), LASSO will select only one feature from a group of correlated features, usually arbitrarily. One might consider "group selection" instead;
3. even in the $n > p$ case, for correlated features, Ridge regression exhibits better predictive power than LASSO, although Ridge regression will not help feature selection and the model is not very interpretable.

8.4 Selection Versus Shrinkage

In the previous Sects. 8.2 and 8.3, the regularization/penalization techniques (LASSO and Ridge) are used in linear regression models. The techniques work generally in two different ways, by selecting variables and by shrinkage of the coefficients in a regression model. Variable selection means choosing a subset from the feature set as predictors to solve the problem. On the other hand, shrinkage means reducing the estimated coefficients in the regression, possibly to 0 in some cases. The corresponding variable(s) can be left out. Thus, shrinkage can also be regarded as a kind of dimensionality reduction method.

Example 8.3 As discussed in Sect. 8.3, LASSO uses a ℓ_1-norm penalty in selecting variables while Ridge uses a ℓ_2-norm penalty and results in coefficients shrinkage. Figure 8.2 (*left*) shows the variables recommended for different values of λ values. If the penalty term is larger, fewer variables have nonzero coefficients. LASSO will suggest the number of key variables from 12 to 3. In contrast, Ridge regression still recommends the same number of 14 variables for a different penalty λ. The strengths of the techniques are summarized in Table 8.2.   □

In general, the ℓ_1-penalty and ℓ_2-penalty techniques can be utilized with different loss functions (defined in Chap. 2.) The Ridge penalty will be of the form

$$L(\beta) = f(\mathbf{Y}, \mathbf{X}, \beta) + \lambda \sum_{i=1}^{p} \beta_i^2$$

Table 8.2 Summary of various variable selection methods via regression

Methods	Description	Strength
Linear regression	Obtaining a linear combination of input variables such that the values are close to the observed response via $(\mathbf{X'X})^{-1}\mathbf{X'Y}$	Predicts outcomes
Ridge	Utilizes ℓ_2-norm penalty to estimate the coefficients while direct inverse of $\mathbf{X'X}$ is not available	Estimates coefficients
LASSO	Utilizes ℓ_1-norm penalty to reduce some coefficients to exactly 0	Selects key variables

and the LASSO penalty will be of the form

$$L(\beta) = f(\mathbf{Y}, \mathbf{X}, \beta) + \lambda \sum_{i=1}^{p} |\beta_i|$$

where f is the loss function based on the data, and β_i are its parameters. By providing such a penalty, the statistical learning model will use a small number of input variables to learn the underlying relationship. Generally speaking, LASSO will help in selecting variables while Ridge will exhibit better predictive power.

Two main advantages of using a selection and shrinkage method are

- Prediction accuracy
  Reducing model complexity results in reducing the variance at the cost of introducing more bias. If the sweet spot (minimal) of the total error (variance and the square of bias) can be obtained, one can improve the model's prediction;
- Model interpretability
  With too many predictors, it is hard to grasp all the relations between the variables. By sacrificing some accuracy, a simpler model can help in getting the big picture. (Interpretability is further discussed in Sect. 10.4.)

Besides penalization, many other techniques can help in selecting variables and reducing dimension. Some typical examples are listed next.

Example 8.4 (Correlations in the Body Fat Dataset) One naive method of selecting contributing input variables is studying the correlation between the response and the input variables and then selecting those input variables whose correlation is far from zero. Table 8.3 shows the correlations between the BodyFat and other variables. If one uses 0.7 as a hard threshold, three variables will be selected: BMI, Chest, and Abdomen. By contrast, linear regression via LASSO leads to Age, Height, and Abdomen.   □

Example 8.5 (Variance Inflation Factor (VIF) in the BodyFat Dataset) Three variables can be selected: BMI, Chest, and Abdomen as shown in Example 8.4. However, the selected variables are also highly correlated, as can be seen in

Table 8.3 Correlations between the response and the input variables in the BodyFat dataset

Variable	Correlation	Variable	Correlation
Age	0.29	Hip	0.63
Weight	0.61	Thigh	0.56
Height	−0.09	Knee	0.51
BMI	0.73	Ankle	0.27
Neck	0.49	Biceps	0.49
Chest	0.70	Forearm	0.36
Abdomen	0.81	Wrist	0.35

Table 8.4 Correlations between BMI, chest, and abdomen in the bodyFat dataset

	BMI	Chest	Abdomen
BMI	1	0.91	0.92
Chest	0.91	1	0.92
Abdomen	0.92	0.92	1

Table 8.5 VIF index of the input variables in the BodyFat dataset

Variable	VIF	Variable	VIF
Age	2.25	Hip	15.16
Weight	33.79	Thigh	7.96
Height	2.26	Knee	4.83
BMI	16.16	Ankle	1.95
Neck	4.43	Biceps	3.67
Chest	10.68	Forearm	2.19
Abdomen	13.35	Wrist	3.38

Table 8.4, the so-called *multicollinearity*. One might not want to include all three variables in a model because just one variable might represent the other two.

The variance inflation factor (VIF) quantifies the severity of multicollinearity under the scenario of linear regression. It provides an index that measures how much the predictor variable can be replaced by the other variables in the model. The key idea is to regress the predictor variable on other input variables. If all 14 predictor variables in the Body fat data are considered, the VIF index is shown in Table 8.5. A rule of thumb for VIF is to remove the variables with index higher than 10. The variables, Weight, BMI, Chest, Abdomen, and Hip should then be removed. □

8.5 Further Refinements

Elastic Net is a regularization technique for variable selection and a hybrid of LASSO and Ridge regression. The extension adds a *penalty function* that can be estimated as a weighted average in the ℓ_1 and ℓ_2 norms as

$$L(\beta) = \|(\mathbf{Y} - \mathbf{X}\beta)\|_2^2 + \lambda_1 \sum_{i=1}^{p} \beta_i^2 + \lambda_2 \sum_{i=1}^{p} |\beta_i|$$

where λ_1 and λ_2 are constant weights to be estimated. The equivalent expression

$$L(\beta) = \frac{1}{n}\|(\mathbf{Y} - \mathbf{X}\beta)\|_2^2 + \lambda((1 - \alpha) \sum_{i=1}^{p} \beta_i^2/2 + \alpha \sum_{i=1}^{p} |\beta_i|)$$

is more common. Clearly, LASSO regression ($\alpha = 1$) and Ridge regression ($\alpha = 0$) are extreme particular cases.

The main advantage of using an Elastic Net solution is that it simultaneously does automatic variable selection and continuous shrinkage. It can select groups of correlated variables and it often outperforms both LASSO and Ridge regression. The Elastic Net solution has been successfully applied to various studies of microarray datasets of high dimensionality (many thousands of predictor genes, large p) and often a small sample size (small n.) Typically, these genes are likely to share the same biological "pathway", which in turn causes them to exhibit correlations. An ideal variable (gene) selection method should a) eliminate the trivial genes, and b) automatically include whole groups into the model once one gene among them is selected. By comparison, LASSO regression lacks the ability to reveal the grouping information because if there are groups of highly correlated variables, LASSO tends to arbitrarily select only one from each group. Consequently, these selected models are difficult to interpret because covariates that are strongly associated with the outcome are not included in the predictive model. (Further details can be found in the literature, for example, [2–4].)

References

1. Hoerl, A. E., & Kennard, R. W. (1970). Ridge regression: Biased estimation for nonorthogonal problems. *Technometrics, 12*(1), 55–67.
2. Segal, M. R., Dahlquist, K. D., & Conklin, B. R. (2003). Regression approaches for microarray data analysis. *Journal of Computational Biology, 10*(6), 961–980.
3. Tibshirani, R. (1996). Regression shrinkage and selection via the lasso. *Journal of the Royal Statistical Society: Series B (Methodological), 58*(1), 267–288.
4. Zou, H., & Hastie, T. (2005). Regularization and variable selection via the elastic net. *Journal of the Royal Statistical Society: Series B (Statistical Methodology), 67*(2), 301–320.

Chapter 9
Machine Learning Approaches

Deepak Venugopal and Max Garzon (ID)

Abstract Machine learning algorithms can train a model to extract some hidden patterns in a dataset to solve a problem or elucidate dependencies among the predictors and thus select or extract features that enable solutions to complex questions from large datasets. This chapter reviews various machine learning methods for dimensionality reduction, including autoencoders, neural networks themselves, and other methods.

9.1 Autoassociative Feature Encoders

Autoencoders are neural networks (defined in Sect. 2.2) that learn to decompose the identity function $I(\mathbf{x}) = \mathbf{x}$ as a composite of an encoding function f and its inverse f^{-1} that decodes its input back to reconstruct the original data $\mathbf{x} = f^{-1}(f(\mathbf{x}))$. Autoencoders perform dimensionality reduction if the encoding representation is constrained to be of lower dimension than the dimensionality of the inputs. In general, autoencoders work in unsupervised machine learning mode (described in Sect. 2.3) since no labels are expected in the data. In other words, they employ *self-supervision* since their goal is to reconstruct the given input. This section describes the general structure of an autoencoder and illustrates various types using the MNIST digits dataset (described in Sects. 5.1 and 11.5) as a running example to reveal their operation and assess their performance.

Example 9.1 For the classification problem of [**CharRC**] and the MNIST dataset consisting of images of handwritten digits (discussed in Sects. 2.2 and 11.5), a solution classifier must place each digit image into one of 10 classes (one of the digits $0, \ldots, 9$). An autoencoder solution might rather proceed as illustrated in Fig. 9.1. It must learn to extract low-level primitives (e.g., curves, lines, shapes) in any variant of a specific input image that are informative enough to be able to

D. Venugopal (✉) · M. Garzon
Computer Science, The University of Memphis, Memphis, TN, USA
e-mail: dvngopal@memphis.edu; mgarzon@memphis.edu

© The Author(s), under exclusive license to Springer Nature Switzerland AG 2022   179
M. Garzon et al. (eds.), *Dimensionality Reduction in Data Science*,
https://doi.org/10.1007/978-3-031-05371-9_9

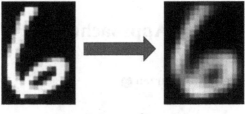

Autoencoder

Fig. 9.1 An autoencoder for the MNIST dataset might produce an encoding (*not shown*) of the input image for a 6 (*left*) and the reconstruction of the image from its encoding (*right*)

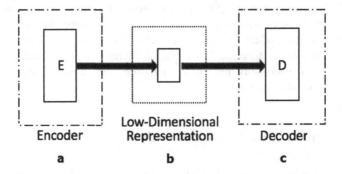

Fig. 9.2 Workflow of a typical autoencoder: (**a**) encoder E that is constrained to reduce into a (**b**) lower dimensional representation of the input, and (**c**) decoder D that reconstructs the original input

reconstruct it from them, as opposed to just decide which digit it is. For instance, a classifier could distinguish between an image of an 8 from the image 7 simply using the pixels in the image to check if the lines are straight or curved. On the other hand, to an autoencoder that needs to learn to retain fewer key additional features to reconstruct both handwritten 7s and 8s, how a digit is written (e.g., whether it has intersecting curves or there are two distinct convex shapes that are joined at a point, or the like) is useful. Thus, a reconstruction ability forces the model to learn a representation that focuses on deeper features of the same inputs, but with lower degrees of freedom. In other words, an autoencoder exhibits deeper richer semantic meaning (one might even say knowledge).   □

Figure 9.2 shows the three main components in the typical workflow for an autoencoder algorithm:

- The encoder E learns an encoding as a lower dimensional representation from the inputs; the corresponding function g is parameterized by the weights of a neural network, say ϕ, i.e.,

$$\mathbf{z} = g_\phi(\mathbf{x}).$$

- Likewise, the decoder D takes as input the lower dimensional vectors z generated by the encoder E and learns to reconstruct the original data from the lower dimensional representations $\mathbf{z}$. The corresponding function f learned by the decoder is parameterized by its own weights θ, i.e.,

$$\mathbf{x}' = f_\theta(z) = f_\theta(g_\phi(\mathbf{x})),$$

The loss function in the autoencoder is measured by the difference between the input $\mathbf{x}$ and its reconstruction $f_\theta(g_\phi(\mathbf{x}))$. By design, if the autoencoder can reproduce the exact same input every time, i.e., if $f_\theta(g_\phi(\mathbf{x})) = \mathbf{x}$, then it will not be particularly useful for dimensionality reduction.

Several approaches have been developed that constrain the autoencoder to learn an encoding in a reduced dimension, including

- *Undercomplete Autoencoders* have a smaller number of hidden units to encode the input.
- *Sparse Autoencoders* add a sparsity penalty to perform dimensionality reduction.
- *Variational Autoencoders* perform dimensionality reduction by representing the input with latent variables, each being assumed to represent a normal distribution.

9.1.1   Undercomplete Autoencoders

Example 9.2 Figure 9.3 illustrates an undercomplete autoencoder. In this architecture, the input and output have the same number of units, but the number of units in

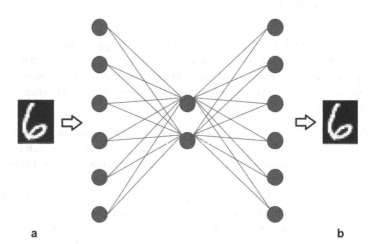

a   b

Fig. 9.3 Schematic representation of an undercomplete autoencoder (**a**) an input and (**b**) its reconstructed output

the hidden layer is much smaller and forces it to combine these features to generate *meta-features* that can be decoded back to reproduce (a close approximation of) the same input.   □

The most common loss function in use is the mean squared error

$$L(\mathbf{x}, f_\theta(g_\phi(\mathbf{x}))) = \frac{1}{2n} \sum_{i=1}^{n} ||\mathbf{x}_i - f_\theta(g_\phi(\mathbf{x}_i))||^2,$$

where $\mathbf{x}_i$ is the ith training instance and $||\mathbf{x}_i - f_\theta(g_\phi(\mathbf{x}_i))||^2$ is the squared error between the input and reconstructed output. Undercomplete encoders have a direct relationship to PCA (described in Sect. 4.1). If the encoder function f_θ and the decoder function g_ϕ are constrained to be linear functions and the loss function is the summed mean squared error loss, then it can be shown that autoencoders are equivalent to PCA, i.e., the optimal dimensions in g for the autoencoder correspond to the principal components learned by PCA [7]. In general, undercomplete autoencoders can learn nonlinear manifolds, i.e., they are capable of representing data with nonlinear surfaces since f_θ and g_ϕ are typically nonlinear. Compared to approaches such as PCA, this increased flexibility may also result in the autoencoder not learning the underlying input features but simply copying input to output. For example, if the encoder has too many layers, the preceding layers increase the capacity of the autoencoder making it more likely to memorize inputs rather than extract latent features for effective dimensionality reduction, even if the output layer's dimension is small.

9.1.2 Sparse Autoencoders

Example 9.3 Figure 9.4 shows the architecture for a sparse autoencoder. In this architecture, the input, output, and hidden layers have the same number of units. Learning constrains a percentage of nodes in the hidden layer so that these units are not active, i.e., they produce an output of 0. This forces the autoencoder to learn a reduced dimensional representation for the input.   □

Undercomplete autoencoders perform dimensionality reduction by reducing the number of units in the layer corresponding to the encoder's output which forces the neural network to represent the input with a smaller dimension vector. On the other hand, sparse autoencoders reduce the number of active units in the encoder (a unit in a neural network is *active* when its output is close to 1, as described in Sect. 2.2). If only a small number of units are allowed to be active to encode given inputs, then the network is forced to only encode features that are important to reconstruct the inputs and leave out irrelevant features. A sparsity penalty constrains the activated units for an input, i.e., the loss function in the autoencoder includes a penalty term on the encoder output given by

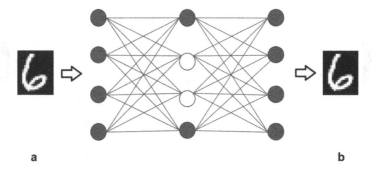

a   **b**

Fig. 9.4 Schematic representation of a sparse autoencoder for (**a**) an input and (**b**) its reconstructed output. The nonfilled nodes represent the units that are not activated (i.e., their output is 0)

$$L(\mathbf{x}, f_\theta(g_\phi(\mathbf{x})) + \Sigma(h) ,$$

where $\Sigma(h)$ is a regularization penalty imposed on the representation generated by the encoder. Similar to Lasso regularization (described in Sect. 8.3), it can be shown that the penalty term is a ℓ_1-penalty over the encoder output.

$$\Sigma(h) = \lambda \sum_{i=1}^{k} |h_i| ,$$

where λ is a hyper-parameter that controls the number of activations and $|h_i|$ is the absolute value of the output from the ith unit in the encoder. Thus, if the encoder is able to encode the input with just a few activated units, the penalty term is minimized. A slightly modified version of backpropagation is used to learn weights through gradient descent for this modified loss function. (More details can be found in [1].)

9.1.3 Variational Autoencoders

One of the most popular types of autoencoders is the variational autoencoder (VAE) [3]. VAEs are a kind of *generative* model in machine learning.

Example 9.4 Figure 9.5 shows an example of a variational autoencoder (VAE). The VAE encoder converts the inputs into *latent vectors* (comparable to the latent features arising in nonnegative matrix factorization (NMF) in Sect. 4.3). Each latent vector is a kind of code that represents the input. To do this conversion, the encoder learns the parameters (mean and standard deviation) for a normal distribution over the latent vectors. The decoder then samples a latent vector from this normal distribution and then reconstructs the output.   □

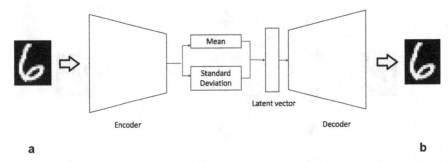

a   b

Fig. 9.5 Schematic representation of a variational autoencoder for (**a**) an input and (**b**) its reconstructed output

In general, VAEs aim to model the data distribution. For instance, if $P_{\mathbf{x}}$ is a probabilistic model of the data distribution over images in the MINST dataset (each data point $\mathbf{x}$ is a matrix of pixels), then the model should assign high probabilities when $\mathbf{x}$ is an arrangement of pixels that appears to be a natural image. Thus, a VAE can generate as many samples as needed from the distribution $P_{\mathbf{x}}$, where the generated samples are similar (but not identical) to instances observed in the data. In the case of graphical data, they could be images that look like real images but are in fact synthetic. This makes VAEs very powerful since they essentially provide unlimited amounts to data. This is particularly useful in applications where it may be hard to obtain or display real data. For instance, to preserve privacy, fake images of people can be generated by VAEs, of student data that do not correspond to real students, or of health records that do not belong to any real person.

VAEs are derived based on a well-known technique in probabilistic models called *variational inference*. In variational inference, the idea is to approximate a data distribution that is unknown by a probability distribution from a known distribution family (e.g., Gaussian). The distribution from the known family is then estimated from the data, typically using optimization methods. VAEs apply this idea to autoencoders. In this sense, VAEs are quite different from other autoencoders since the ultimate goal in VAEs is not really dimensionality reduction but rather data generation. Dimensionality reduction is a side bonus of learning a generative model for the data.

For the MNIST dataset, a generative model for decimal digits 0–9 can be developed as follows. To generate a certain image of $\mathbf{x}$, the model sets pixel values in the image such that the image resembles the shape of the digit. The model first samples $\mathbf{x}$ to get a set $\mathbf{z}$, which will be the latent variable, and then conditions on z-values for pixels to generate the image. For different z's, the sample generates different handwritten digits. The latent variable may need to capture richer information to be able to generate all possible variants of a digit. For example, apart from the digit to generate, the latent information should include the angle at which the digit is written, the thickness of the writing, curvature, and so forth. Capturing this information is not straightforward, unless the distribution over the latent variables itself is complex. Instead, VAEs assume a simple distribution to generate z and then pass this through a complex function over several layers in the autoencoder to map the latent variables to an output. In particular, VAEs assume that each dimension in the latent representation is independent of other dimensions and is normally distributed, so that it can be easily sampled. The latent variable represents the dimensionality-reduced instances for the data distribution.

If z was randomly sampled from a normal distribution and then an image was generated from $P_X(\mathbf{x} \mid z)$, this probability may be close to 0 in most cases; in other words, the latent representation does not correspond to an actual meaningful digit image. The main idea in VAEs is to use a variational distribution Q from which the latent variables are sampled. That is, Q is a distribution over the latent variables, and sampling a value from this distribution results in an assignment to the latent variables that corresponds to a valid image of a digit. In other words, given an input image $\mathbf{x}$, Q should generate a z that produced $\mathbf{x}$. As in standard variational inference, the objective is to minimize the distance (measured using a standard measure such as the Kullback–Leibler or KL divergence, as described in Sect. 5.2) between the variational distribution and the true distribution. The KL divergence can be derived as

$$\mathscr{D}(Q(z)\|P(z|X)) = E_{z\sim Q}[log\, Q(z) - log\, P(X|z) - log\, P(z)] + log\, P(X),$$

where $z \sim Q$ indicates that the expected value is computed using samples from the distribution Q. This formula can also be written as

$$log\, P(X) - \mathscr{D}(Q(z)\|P(z|X)) = \mathbb{E}_{z\sim Q}[log\, P(X|z) - \mathscr{D}(Q(z|X)\|P(z)].$$

The left-hand side represents the data distribution P_X and a term that becomes smaller if Q can generate the latent variables z that can be used to generate a valid data instance (say, an image). The right-hand side has two parts, Q and P. Q encodes a data point $\mathbf{x}$ to generate zs and P takes as input z to generate $\mathbf{x}$. This can be mapped into the autoencoder framework where the encoder generates values of latent features for the decoder to be able to generate a valid output. An analogous interpretation is that the encoder generates a code of reduced dimensional for the input and the decoder can generate the output given this code.

To learn VAEs using BackProp (defined in Sect. 2.2), a stochastic layer is required. Specifically, given an input, the encoder samples each latent variable using a normal distribution. However, it turns out that technically doing this is infeasible since BackProp is based on gradient descent because it is impossible to take derivatives of stochastic variables. Therefore, VAEs introduced an additional idea called the *reparameterization* trick. The main idea here is to make the sample for the latent variables independent of the normal distribution parameters. Specifically, if a unit corresponding to the latent variable has to sample from the normal distribution $N(\mu(\mathbf{X}), \Sigma(\mathbf{X}))$, it first generates a sample ε from $N(0, I)$, i.e., a normal distribution with mean 0 and covariance matrix the identity matrix. The actual value of the latent variable z is then deterministically computed using ε along with the parameters of the original normal distribution μ, Σ as

$$z = \mu(\mathbf{X}) + \Sigma^{1/2}(\mathbf{X}) * \varepsilon.$$

The reparameterization trick allows the gradient computations to be performed over the layer encoding the latent variables. BackProp can therefore be applied indeed to learn the parameters of the VAE network.

Though VAEs provide a sound probabilistic framework for autoencoders, there are some practical concerns. One of the well-known difficulties with using VAEs is that the latent codes generated tend to be uninformative. Specifically, it is hard to understand what exactly each of the latent variables does. For the handwritten digits in the MNIST dataset for example, ideally latent variables should encode specific peculiarities of the writing, but this is hard to enforce in a VAE. Devising methods to learn disentangled representations from VAEs that explain the latent variables remains an active area of research [10].

9.1.4   Dimensionality Reduction in MNIST Images

This subsection presents some experimental results on dimensionality reduction using autoencoders on the MNIST dataset (described in detail in Sect. 11.5).

The autoencoder architecture for MNIST is as follows. Since the data consists of images, convolutional neural nets [9] are used in the encoder and decoder since they are the de facto neural networks for image processing. Each convolutional layer has kernels to perform convolutions (i.e., dot products) with the input to extract features. The standard encoder architecture consists of three convolutional layers, with each convolutional layer followed by a pooling layer that reduces the size of the input by downsampling it. Similarly, the decoder consists of three convolutional layers and each convolutional layer is followed by an upsampling layer that increases the size of the input.

The output from the encoder layers is a 128D vector, a substantial reduction of the 784 dimensions in the original image. Table 9.1 shows the decreasing loss,

Table 9.1 Performance of autoencoders for dimensionality reduction of the MNIST dataset for various reduced dimensions in the output layer (*columns 2–4*). The reconstruction loss (a measure of similarity between the reconstructed images and the original images) decreases toward 0 as the number of epochs increases

Epochs	Loss with 128D	Loss with 64D	Loss with 32D
5	0.110	0.141	0.177
10	0.100	0.131	0.161
15	0.099	0.126	0.153
20	0.096	0.122	0.150
25	0.094	0.118	0.149

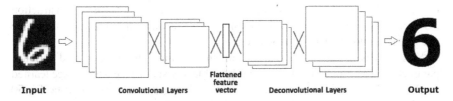

Input   **Convolutional Layers**   Flattened feature vector   **Deconvolutional Layers**   **Output**

Fig. 9.6 The architecture of a convolutional neural network (CNN) consist of three layers: convolutional layers (*left*) that look for spatiotemporal correlations in the input features and produce a decomposition of the input into a flattened feature vector (*middle*), and deconvolutional layers (*right*) that reconstruct the input from the flattened feature vector into a high-level representation

Table 9.2 Reconstruction and Kullback–Leibler (KL) divergence loss for VAEs on the MNIST dataset

Epochs	Reconstruction loss	KL loss
1	210	3.30
10	149	6.15
20	145	6.53
30	144	6.40
40	142	6.58
50	142	6.60

i.e., a measure of the difference between the original images in the dataset and the corresponding reconstructed images. The loss value decreases toward 0 as the number of epochs (iterations through the dataset) is increased. Furthermore, the dimensionality of the encoder output also affects the loss.

The VAE architecture for MNIST digit generation (Fig. 9.6) consists of 2 convolutional layers and then a fully connected layer of 16 units, i.e., each output of the convolutional layer is connected to all 16 units in the fully connected layer. The fully connected layer is connected to the sampling layer for the latent variables. The decoder takes as input the sampled code and uses one fully connected followed by two convolutional layers to reconstruct the image. The reconstruction and the KL losses are shown in Table 9.2. The reconstruction loss measures how well the input image can be reconstructed from the encoded image, whereas the KL loss measured the difference between the prior probability on the latent space $P_x(z)$ and $Q(z \mid \mathbf{x})$, i.e., the distribution that encodes probabilities in the latent space given that the latent representation does correspond to a digit image. The KL loss reflects the spreading of the latent representations of the digit images in the latent space. Thus, the decoder can generate an image for a given random point in this latent space. This gives the VAE the ability to generate new digit images by randomly sampling a point from the latent space. The reconstruction loss ensures that points that are close to each other in the latent space actually correspond to similar images, and therefore reconstructing them is easy for the decoder. The combination of the two losses means that the latent space represents clusters corresponding to different digits and that there is a smooth interpolation between these clusters. Table 9.2 shows the reconstruction and the KL loss for different numbers of epochs. As clusters begin to form in the latent space corresponding to the different digits, the reconstruction loss becomes smaller and the KL loss increases since the digit representations are not spread randomly across the whole space.

Since VAEs represent the reduced dimensions through the mean and variance parameters in a normal distribution, it is possible to visualize the latent space learned by the VAE for the data points, as illustrated in Fig. 9.7. The latent space for 50 epochs shows that digit images corresponding to different digits are in different clusters. Furthermore, there is a smooth interpolation across clusters. For example, the representations for the digit 8 are close to those for 9 since the shape of 8 and 9 is such that an 8 can be obtained from a 9 by adding a single arc. This type of interpolation is extremely useful since a new sample from the latent space generates a new image that was never present in the training data. This property makes VAEs a really powerful generative model.

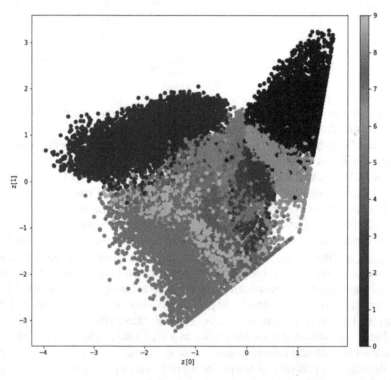

Fig. 9.7 Ten clusters distinguishable in the latent space generated by the VAE for MNIST handwritten digits show that codes generated for similar digits are also similar in the latent space, from which VAE reconstructs the inputs

9.2 Neural Feature Selection

In this section, neural networks are shown to facilitate DR in a manner akin to DR using Shannon entropy discussed in Chap. 6. The key idea is that a feature X_i is *neurally dependent* on a set of features F if its values can be predicted by a feedforward neural net using the remaining features of a data point as inputs. Variations of this idea are explored in detail using an application to develop a neurocontrol for emotional displays in avatars, specifically for AutoTutor (https://en.wikipedia.org/wiki/AutoTutor), following up on [6].

Anthropomorphic representations of software agents (avatars) can be used (e.g., instead of static pictures) in order to enhance the quality of communication and remote interaction among humans or even between computer and human users. Notable examples are AutoTutor, a software agent capable of tutoring a human user in a restricted domain of expertise, such as computer literacy or elementary physics, at the level of an untrained human tutor; Grace (www.cmu.edu/cmnews/020906/020906_grace.html), a robot that registered and delivered a speech at the 2002 AAAI conference on Artificial Intelligence; and a videophone model for videoconferencing on low bandwidth channels [13].

Example 9.5 AutoTutor is an embodied conversational agent consisting of a dialog module that handles the computational intelligence to carry on a conversation with a human student and an interface module embodied in a talking head to convey nonverbal feedback [5, 13]. The simplest method to design such an interface consists of designing an ontology of prototypical facial expressions or animations that presumably reflect the desired type of cartoon-like responses by the avatar to answers in natural language given by a student to questions posed by the avatar in a round of conversation to learn a certain concept, based on pre-programmed action sequences. While the result may suffice in some cases, it is clear that the resulting solutions suffer from a number of problems, such unnaturalness of the expressions, robotic appearance, inappropriate responses, and perhaps worst of all, a continuing programming effort to produce scripts that do not adapt to a rich variety of circumstances in a free-wheeling interaction with humans.   □

Another example is the videophone model described in [13]. It introduced a technique that increases the efficiency of the communication at least 1000-fold by replacing unrealistic video transmissions with a three-stage process, namely extraction and coding of facial features into text form, transmission over a low-bandwidth channel, and reconstruction of the visual expression in synch with voice at the receiving end.

By contrast, recurrent neural nets have been trained as neurocontrollers for emotional displays in avatars, on a continuous scale of positive, neutral, and negative feedback, that are meaningful to users in the context of tutoring sessions on a particular domain (e.g., computer literacy) for AutoTutor. It is desirable that the avatar interacts with the live user in a naturalistic fashion, so that the conversation takes place in a nonprescribed manner and in real time. The DR techniques using neural nets are described below to show a complete application for the design of such an agent, Artsie, that autonomously and dynamically generates and synchronizes the movements of facial features such as lips, eyes, eyebrows, and gaze in order to produce facial displays that convey high information content in nonverbal behavior to untrained human users. The neurocontrol is modular and can be easily integrated with semantic processing modules of larger agents that operate in real time, such as videoconference systems, tutoring systems, and more generally, user interfaces coupled with affective computing modules for naturalistic communication. A novel technique, cascade inversion, arises from this DR technique that provides an alternative to backpropagation through time, which may fail to learn recurrent neural nets from previously learned modules playing a role in the solution.

9.2.1 Facial Features, Expressions, and Displays

The human face has the greatest mobility and facial display repertoire among all primates. Nonverbal facial behavior may account for as much as 93% of human affective communication events [2], compared to as low as 7% from words. Blushing, eye rolling, eye winking, and tongue sticking are gestures that people use in daily life to quickly communicate feelings, emotions, and attitudes far more effectively than words. Mehrabian's study [11] further suggests that incongruencies between word and facial displays are resolved mostly by nonverbal means (the study suggested about 7%/45%/55% attribution to verbal/vocal/facial cues, respectively). Facial features are used by human speakers in a highly dynamic and well-orchestrated process that has nothing to envy in the complexity and information bandwidth already present in verbal discourse process. For example, facial animations alone are an old and fertile art for communication, as evidenced by long traditions of popular cartoonists and animators.

A human face is normally endowed with 43 muscles that can be flexed in a continuous range each. The first step in this challenging DR task is thus to choose an appropriate set of representative features that control a static facial expression, i.e., the actual face made in an instant by a human in communication with others, even over the phone. Facial expressions have a structure (akin to following a "grammar") of their own, e.g., they tend to be vertically symmetric and primarily involve the eyes, the mouth, and the eyebrows. Furthermore, these expressions are chained in smooth and continuous *facial displays* that change dynamically in ways peculiar to each individual. A *facial display* is defined as an animation consisting of a sequence of various key static frames (ranging from 5 to 15) shown at appropriate speed to be perceived as motion by the human eye. To address this complexity, Ekman developed a nomenclature to describe them, the Facial Action Coding System (FACS) [4] based on 64 action units (AUs). The first reduction is to judiciously select ten of them, namely two eyebrows (left LEB and right REB), distance between the eyebrows, two eye lids (LEL, REL), two eye directions—horizontal

(HED) and vertical (VED)—as well as a mouth height (MH), mouth width (MW), and a frame duration. This reduced information still permits a continuous and smooth animation, as facial expressions can be interpolated to produce a continuous facial display. Moreover, the FACS system identifies six basic human emotions (i.e., *confused, frustrated, negative neutral, neutral, positive neutral, and enthusiastic*). *Rating* a facial expression or display is the forward classification problem of deciding for a given expression or display what emotion it is conveying or at least how positive (or negative) it is. Neurocontrol design is actually the inverse problem, i.e., given a particular emotion to be conveyed by an avatar (perhaps on a continuous scale), how to configure these features so that a human will evaluate it to the given (degree of) emotion. This degree is referred to as the *emotional attitude* and is simply quantified by a number in the interval [0, 1], with 1 being most positive, 0.5 being neutral, and 0 being very negative.

9.2.2 The Cohn-Kanade Dataset

A training set based on recordings of human facial displays produced by real life professional actors was captured in a well-known labeled database, known as the Cohn-Kanade database [8]. This set consists of approximately 500 image sequences from 100 subjects ranging in age from 18 to 30 years. Sixty-five percent were female; 15 percent were African-American, and three percent were Asian or Latino. Each begins from a neutral or nearly neutral face. For each, an experimenter described and modeled the target display. Six were based on descriptions of prototypic emotions (i.e., joy, surprise, anger, fear, disgust, and sadness). These six tasks and mouth opening in the absence of other action units were annotated by certified FACS coders. Seventeen percent (17%) of the data were comparison annotated. Inter-observer agreement was quantified with Cohen's *kappa coefficient*, i.e., the proportion of agreement above what would be expected to occur by chance (0 is bad and 1 is perfect). The mean kappa for inter-observer agreement was 0.86. Image sequences from neutral to target display were digitized into 640 × 480 or 640 × 490 pixel arrays with 8-bit precision for grayscale values. (Further details can be found at *ri.cmu.edu/project/cohn-kanade-au-coded-facial-expression-database/*) (Fig. 9.8).

9.2.3 Primary and Derived Features

Modules that provide partial solutions to the inverse problem for facial gestures were obtained by backpropagation training (as described in Sect.2.2) based on input features and the rating of the desired emotional attitude. Feedforward nets with two hidden layers [40 10 8 1] were required for learning to converge while training for one feature at a time with rating as input instead. The performance of the trained

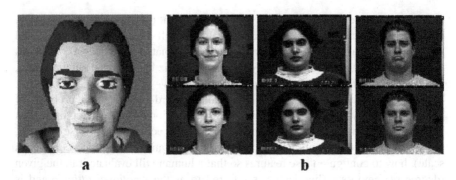

Fig. 9.8 (**a**) An avatar Artsie for AutoTutor. (**b**) Facial expressions (positive: *left*; neutral: *middle*; and negative: *right*) by trained actors in the Cohn-Kanade dataset

Table 9.3 Most selected facial features are derivable, except for the (primary) ones marked *

Features	Training (%)	Testing (%)
Rating*	95.0	69.0
Left eyebrow (LEB)	84.0	89.4
Right eyebrow (REB)	88.0	86.5
Distance between eyebrows (D)	98.7	85.5
Left eye lid (LEL)	98.0	91.0
Right eye lid (REL)	97.5	93.2
Horizontal eye direction (HED)	98.5	94.2
Vertical eye direction (VED)	99.0	98.0
Mouth height (MH)*	99.0	97.0
Mouth width (MW)*	99.0	86.0
Time duration (TD)*	99.0	80.0

networks can be seen in Table 9.3. Rating and all sequence features in the frames, except for distance between eyebrows, mouth width, and time duration, were shown to be derivable individually.

The mouth width was again found to be a primary feature that proved capable of deriving the remaining features together with the frame duration and emotional attitude. In particular, the height of the mouth was now, perhaps surprisingly, derivable from these primary features. However, the mouth width again was found to be a primary feature that proved capable of deriving the remaining features together with the frame duration and emotional attitude. This is likely due to a high correlation between the two features. On the other hand, the distance between the eyebrows also turned out to be a primary feature for which a neural network could not be trained, but with which other features could be derived (Fig. 9.9).

In the next step, given the structure of facial expressions and the multiple dependencies among their constituent features, multiple combinations of features were shown to be jointly derivable from the remaining features, all the way up to seven (7) combinations at a time, as shown in Table 9.4. With the corresponding

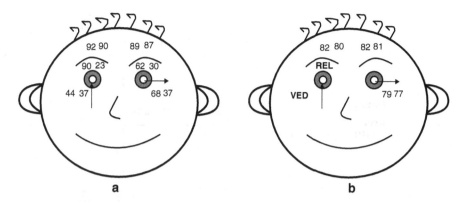

Fig. 9.9 (a) Performance of feedforward neural nets (FNNs) for derivable features for LEB, REB, HED, and VED, shown in the corresponding places (training: left score in black and testing: right score in gray). (b) Likewise for triple jointly derivable features

Table 9.4 Joint derivation of seven (7) features at a time, including left and right eye lid (LEL and REL), horizontal and vertical eye direction (HED and VED), left and right eyebrows (LEB and REB), with the remaining features as inputs (rows)

Features	Training (%)	Testing (%)
Rating*		
Left eyebrow (LEB)	96.0	93.6
Right eyebrow (REB)	97.0	93.5
Mouth height (MH)	99.4	95.2

trained networks, a neurocontrol for facial expressions was successfully trained with excellent, in fact, nearly optimal results for a neurocontroller, as shown in Fig. 9.10. Since the methodology was kept essentially intact in order to ascertain the effect of the quality of the Cohn-Kanade dataset, it is clear that the improvement is due to the more naturalistic data in the new training set. A posteriori, it is thus easy to see why the complexity of the problem would not allow a simple and direct approach to the problem using recurrent backpropagation on the facial displays to converge to a neurocontrol solution.

This application shows that neural network methods can be used for feature selection in DR so as to further enable a solution to a complex problem (inverse problem for facial displays) that could not be had by a direct approach on the full set of features. It illustrates the power of DR to solve a complex problem, where simpler strategies may not work.

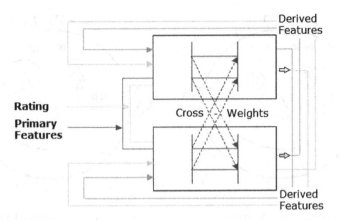

Fig. 9.10 Once the primary features were identified, neural nets can be trained to derive multiple combinations of features (as far as seven at a time) from the rating and the remaining ones; they could then be used as modules for solution to the inverse problem for facial expressions with inputs just the primary features and the rating (emotional attitude) to obtain a full set of features for a facial expressions and eventually for facial displays by iteration over time

9.3   Other Methods

Dimensionality reduction methods typically work in an unsupervised manner, i.e., they learn a lower dimensional representation for the features in the data without considering the response variable. In many cases, however, it may be more effective to reduce the number of features based on the response variables or labels, if available, corresponding in the data points, and therefore use solutions based on supervised learning. Specifically, if the output of the dimensionality reduction method is used within a supervised classifier, then, ignoring the response variable makes dimensionality reduction harder or not efficient, particularly with data that contains a large number of features that are irrelevant to the classification problem. For example, selection of features that are most relevant for classification (as in Chap. 8) may prove more effective in reducing dimensionality. This section presents methods obtained by a commonly used feature selection approach called *wrapper methods*.

Example 9.6 For the character recognition problem [**CharRC**] (described in Sects.2.1 and 11.5), the features are pixels in a certain position in the image, and the selection will choose the most relevant for the classifying of an MNIST image. For, say, an image of the digit 7, the pixels near the top right corner where the two lines intersect are important, while for an image of 4, the pixels at the bottom left corner are more important. Using the labels of the instances to be classified, one can reduce dimensionality by picking only the relevant features for each class.   □

In general, feature selection is a combinatorial problem that is computationally hard to solve since, for example, there are 2^d possible subsets of a set of dimensionality d features to choose from. Naturally, it is infeasible to go over all possible subsets of features if d or the dataset is not small. Two general approaches for feature selection to address this problem are as follows:

- *Wrapper methods* [12] search through the feature space to find the best subset of features using the classifier as a black-box oracle.
- *Embedded methods* integrate the feature selection within the classifier, i.e., the selection is performed directly as part of the learning. Some supervised machine learning algorithms implicitly perform feature selection. Specifically, the decision tree classifier (defined in Sect. 2.2) selects features based on a statistical measure of how well the feature classifies the data. The most popular embedded method for feature selection is LASSO regression that performs ℓ_1 regularization (described in Sect. 8.3). LASSO removes features whose regression coefficients/weights are low or 0.

Wrapper methods for feature selection typically use a search strategy to find the relevant features, with variations depending on the search strategy used to select features for testing.

Example 9.7 Figure 9.11 shows the results of the wrapper method for selecting features in the MNIST dataset. The idea is to use a classifier on different subsets of features and evaluate the performance on these subsets. Only features with the best performance for a given classifier are then selected as representative features, thus resulting in a lower dimensional representation.   □

Different strategies have been developed to select features. *Forward elimination* is an approach that starts with an empty set of features and adds the best feature from the remaining features incrementally. To determine the best feature, the accuracy of the classifier is estimated when a feature is added to it. Several approaches have been used to assess this method, including standard methods such as cross-validation and criteria such as AIC and BIC (as described in Sect. 2.4).

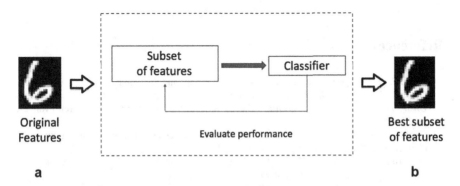

Fig. 9.11 Search of features subsets by typical wrapper methods for (**a**) inputs from on the MNIST dataset and (**b**) corresponding outputs

Table 9.5 Selected features in Recursive Feature Elimination for MNIST for various numbers of features selected (d)

d	SVM	AdaBoost	Decision trees
2	0.28	0.26	0.34
4	0.42	0.46	0.55
8	0.81	0.71	0.81
16	0.87	0.76	0.87
64	0.96	0.85	0.87

Backward elimination uses the opposite strategy. It starts by selecting all the features and then removes the worst feature in successive steps. *Recursive Feature Elimination* (RFE) [12] is a popular variant of this wrapper method that recursively eliminates features using a greedy strategy. It assumes that the classifier has weights that can be used to rank features by their importance. They can be coefficients for features learned in a linear classifier or in SVM (both are described in Sect. 2.2). From these, the least important feature is dropped and the model is retrained using the remaining features. Furthermore, in contrast with other dimensionality reduction approaches, feature selection is more intuitively interpretable since the retained features can be manually validated to see if they are indeed relevant to the classification task.

RFE using support vector machine classifiers can yield impressive results even with a small number of features. Table 9.5 illustrates the results for MNIST classification using RFE for feature selection based on measuring the average F1-score using fivefold cross-validation (described in Sect. 2.4) with various selection sizes, i.e., for each classifier, RFE is used to select d features and use only them to train solutions and compute their cross-validation scores. The quality of the choices is assessed using accuracy across several different supervised learning algorithms (discussed in Sect. 2.2). These results show that, as one increases the number of features that are selected by RFE, the classification accuracy improves for all classifiers, an indication that RFE indeed selects the most relevant features for classification.

References

1. Baldi, P. (2012). Autoencoders, unsupervised learning, and deep architectures. In I. Guyon, G. Dror, V. Lemaire, G. Taylor & D. Silver (Eds.), *Proceedings of ICML Workshop on Unsupervised and Transfer Learning*. Proceedings of Machine Learning Research (Vol. 27, pp. 37–49).
2. Carmel, D., & Bentin, S. (2002). Domain specificity versus expertise: Factors influencing distinct processing of faces. *Cognition, 83*, 1–29.
3. Diederik, P. K., & Max, W. (2014). Auto-encoding variational bayes. In *International Conference on Learning Representations, ICLR*.
4. Ekman, P. (2008). An argument for basic emotions. In *Basic emotions, rationality, and folk theory* (pp. 169–200). Springer.

5. Garzon, M., Drumwright, E., & Rajaya, K. (2002). Training a neurocontrol for talking heads. In *Proceedings of the 2002 International Joint Conference on Neural Networks. IJCNN'02* (pp. 2449–2453). IEEE.

6. Garzon, M., & Sivakumar, B. (2010). A neurocontrol for automatic reconstruction of facial displays. In *Proceedings of the IEEE World Congress on Computational Intelligence-WCCI-10* (pp. 267–272).

7. Hinton, G. E., & Salakhutdinov, R. R. (2006). Reducing the dimensionality of data with neural networks. *Science, 313*(5786), 504–507.

8. Kanade, T., Cohn, J. F., & Tian, Y. (2000). Comprehensive database for facial expression analysis. In *Proceedings of the 4th IEEE Conference on Automatic Face and Gesture Recognition* (pp. 46–53).

9. Krizhevsky, A., Sutskever, I., & Hinton, G. E. (2012). ImageNet classification with deep convolutional neural networks. In F. Pereira, C. J. C. Burges, L. Bottou & K. Q. Weinberger (Eds.), *Advances in neural information processing systems* (Vol. 25). Curran Associates.

10. Mathieu, E., Rainforth, T., Siddharth, N., & Teh, Y. W. (2019). Disentangling disentanglement in variational autoencoders. In K. Chaudhuri, R. Salakhutdinov (Eds.), *Proceedings of the 36th International Conference on Machine Learning*. Proceedings of Machine Learning Research (Vol. 97, pp. 4402–4412). PMLR.

11. Mehrabian, A. (2007). *Nonverbal communication*. Taylor and Francis.

12. Ron, K., & George, H. J. (1997). Wrappers for feature subset selection. *Artificial Intelligence, 97*(1–2), 273–324.

13. Yang, X., Garzon, M. H., & Nolen, M. (2006). Using talking heads for real-time virtual videophone in wireless networks. *IEEE MultiMedia, 13*, 78–84.

Chapter 10
Metaheuristics of DR Methods

Deepak Venugopal, Max Garzon (ID)**, Nirman Kumar** (ID)**, Ching-Chi Yang** (ID)**, and Lih-Yuan Deng** (ID)

Abstract This chapter synthesizes key heuristics distilled from a number of methods that can be applied to dimensionality reduction, leveraging choices such as feature grouping and domain knowledge, as well as the meta-implications of feature selection, such as explainability. Also, some points for reflection on the inherent limitations of dimensionality reduction methods are considered.

10.1 Exploiting Feature Grouping

The dimensionality reduction techniques and solution models defined in the foregoing chapters work best when the input features belong to a single object in the data ontology and thus have common characteristics. For example, there is an implicit assumption in the use of PCA that all p columns in the dataset **X** have a similar scale and/or similar characteristics, e.g., each column represents gene expression or pixel intensities, but not a mixture of both. If not so, PCA may not perform so well, for example if one variable dominates the variance simply because it is scaled by a huge constant with respect to the others. The obvious "fix" is to rescale the variables in the feature by standard statistical normalization into Z-scores (defined in Sect. 11.1) with mean 0 and unit variance.

Example 10.1 (gPCA for Autonomous Vehicle Data Analysis) Data from autonomous vehicles is inherently multimodal since it can contain image data, sensor data, and behavioral data. Jointly reducing the dimensionality for the full feature set may be difficult since each group can have unique characteristics. In this

---

D. Venugopal (✉) · M. Garzon · N. Kumar
Computer Science, The University of Memphis, Memphis, TN, USA
e-mail: dvngopal@memphis.edu; mgarzon@memphis.edu; nkumar8@memphis.edu

C.-C. Yang · L.-Y. Deng
Mathematical Sciences, The University of Memphis, Memphis, TN, USA
e-mail: cyang3@memphis.edu; lihdeng@memphis.edu

© The Author(s), under exclusive license to Springer Nature Switzerland AG 2022   199
M. Garzon et al. (eds.), *Dimensionality Reduction in Data Science*,
https://doi.org/10.1007/978-3-031-05371-9_10

case, reducing dimension for features that have some commonalities is likely to be effective.   □

In general and in practice, this assumption can be met by using the well-known stratification technique from sampling (described in Sect. 1.3), i.e., the feature set can be subdivided into several groups with similar characteristics within each. One obtains a *grouped* variant $\Psi(\cdot)$ of any DR approach $\Psi(\cdot)$, for example, gPCA for *grouped PCA*. Specifically, let g be the number of desired groups in the study, with each group sharing a common scale and/or a similar characteristic for the features within the group. The columns of $\mathbf{X}$ can then be rearranged as described in Sect. 3.1:

$$\mathbf{X} = [\mathbf{X}_1, \mathbf{X}_2, \ldots, \mathbf{X}_g].$$

After performing $\Psi(\cdot)$ (e.g., PCA) within each group $\mathbf{X}_i, i = 1, 2, \ldots, g$ and finding a reduction (its leading principal components) $\mathbf{X}_d^* = \mathbf{X}_d \mathbf{C}_d$, they are put together to obtain a reduction for all components

$$\mathbf{X}^* = [\mathbf{X}_1^*, \mathbf{X}_2^*, \ldots, \mathbf{X}_g^*] = [\mathbf{X}_1 \mathbf{C}_1, \mathbf{X}_2 \mathbf{C}_2, \ldots, \mathbf{X}_g \mathbf{C}_g],$$

where $\Psi_k(\cdot)$ is a dimension reduction procedure for the kth group. It is even possible to mix in different types of DR $\Psi_k(\cdot)$ for different groups. Grouping is commonly used with PCA in statistics with good results.

In summary, gPCA is a variant of any DR method to address the problem with heterogeneous features in the data by grouping them more homogeneously, reducing dimensionality in each subset of a partition of the features set, and then putting all reduced feature groups together into a reduced feature set.

Advantages of grouped DR include:

- It includes domain knowledge to group wisely and is likely to improve the quality of the DR and solutions therefrom.
- A homogenous set of features reduces the sampling variation, as evidenced by common similar techniques in statistics.
- Grouping can clearly reduce computing costs or bring it to the realm of the feasible where the original dimensionality p is large. For PCA for example, it would be inefficient or impossible to find the eigenvalues/eigenvectors for a large variance–covariance matrix $\mathbf{S}$ of size $p \times p$.

10.2 Exploiting Domain Knowledge

Knowledge is a critical component that determines the success of a data science application. The field of knowledge representation (KR) has a rich history starting with its study in philosophy and logic in western philosophy since ancient times. In modern cognitive science and AI, KR is a well-known branch of research that deals with representation of knowledge and using it for problem solving. Well-known KR systems that have been developed include Cyc [11] (for common sense knowledge), WordNet [13] (for language semantics), and MYCIN [28] (an expert system for medical diagnosis). More recently, a system called NELL [20] developed at Carnegie Mellon University crawls web pages to extract data about knowledge from the web and currently contains over 50 million facts. While the role of domain knowledge (DK) in data science is acknowledged by all practitioners, a proper and general definition of what "knowledge" really is remains a fuzzy but essential unresolved issue for all sciences. It is not unreasonable to say that this roadblock has stumped genuine advances in all sciences. This section proposes a general definition for the concept of knowledge in data science and then presents an overview of different approaches that have been used to incorporate domain knowledge into dimensionality reduction methods.

10.2.1 What Is Domain Knowledge?

The definition of what constitutes knowledge has a long history in philosophy, and an entire subfield of *epistemology* has been devoted to study the origin, nature, and other foundational aspects of knowledge since ancient Greek times. A general definition of knowledge is out of scope in this book (and probably overall), so a less ambitious approach is taken here.

Example 10.2 Looking back at the history of science, the *Copernican model* together with *Kepler's laws* of planetary motion is arguably the first piece of scientific knowledge concerning the understanding of our solar system (and eventually the physical universe due to its universality). Subsequent discoveries such as universal gravitation and three laws of physics by Sir Isaac Newton put a crown jewel on this body of knowledge in the domain of physics. A similar statement can be made for discoveries such as the atomic theory of matter and the chemical table of elements in the domain of chemistry, the neuronal model in the domain of neurophysiology, and so forth for other natural science domains.   □

Generalizing these examples, a definition of domain knowledge in data science is proposed as follows.

Domain Knowledge (DK) in data science is (equivalent to) a (set of) model(s) and/or algorithms that can be used to solve various data science problems.

Example 10.3 Most of the models described throughout the foregoing chapters, except perhaps Chaps. 1 and 3, for dimensionality reduction and for problem solving can thus be considered knowledge in the domain of data science. More examples are given in Chap. 11.   □

This general definition implies that knowledge is embedded at every step in the data science pipeline. Specifically, it is used to represent the data through informative features and train the models for these representations to solve specific problems such as classification and clustering, or to make predictions. Further, it implies that domain knowledge is also needed to evaluate and assess the performance of these models in the real world. Based on testing models in the real world, one could potentially discover aspects about the model that were previously unknown, forcing a refinement of the model, thus improving the model over time in an ever going monotonic process to augment the body of knowledge. Domain knowledge thus requires domain *experts* who integrate the models into a coherent and organized body of knowledge and assure their effectiveness, accumulate experience across problems, datasets, and outcomes, and brainstorm about ways to address their shortcomings.

10.2.2 Domain Knowledge for Dimensionality Reduction

Many of the methods for DR explored in previous chapters appear to be domain agnostic, i.e., do not seem to require domain knowledge, except perhaps in the mathematical or computational domains in the form of models and algorithms in data science. It is quite conceivable that a domain expert can see beyond these models and come up with better solutions out of prior experience in the domain. Can Domain Knowledge really help in dimensionality reduction beyond the data science models in Chaps. 2–9 in this book?

Example 10.4 For the problem of malware classification [**MalC**], one can try to answer this question with an experiment, i.e., give the same problem and dataset to an expert in this domain of cybersecurity and also to a proverbial educated layman or data scientist. While the general data scientist can use generic data science models (his domain knowledge), it is more likely that the expert will have more success due to the use of specialized models (domain knowledge) that have historically worked well for this problem in this domain.

For the specific case of dimensionality reduction, domain knowledge can act as a *guide that helps create more impactful representations of the data*, where the impact is measured by its effectiveness in solving the problem of interest. While most of the dimensionality reduction approaches discussed in this book are purely data dependent and do not require external domain knowledge; in practice, the use of domain knowledge can significantly improve the performance of dimensionality reduction. Some existing approaches that embed domain knowledge in dimensionality reduction can be summarized as follows.

Constraint-Based Methods
In real-world data, similarities exist across different data points. Pictures of the same person have a degree of similarity, and health records for the same individual have similarities even if coming from different doctors. Adding this knowledge of similarities in an ontology can help organize the data and produce better dimensionality reduction, e.g., less redundant features. In general, this problem can

be formulated as added constraints on the data instances. Such constraints have also been applied to other unsupervised clustering methods (described in Sect. 2.3). This set of constraints can be represented as a graph, where the nodes are the data instances and the connections are constraints between the instances. In [9] , an approach to dimensionality reduction is described that uses domain knowledge in the form of a graph with link constraints to indicate the degree of similarity of two data instances. The degree of similarity is encoded as a weight. Positive/negative weights on an edge indicate a degree of similarity/dissimilarity between instances, respectively. Dimensionality reduction thus projects the data into a space of lower dimension (like with PCA) that depends on the weights between data instances, say by pushing together similar instances and pushing far apart instances that are dissimilar from one another. Using experimental results in several standard machine learning benchmarks, it has been shown [9] that using this approach with clustering, the reduced features yield better quality clusters compared to using PCA for dimensionality reduction without external knowledge.

Grouping Methods
The constraint-based methods described above add constraints in the form of rows to the data. The same idea can be used for columns. Specifically, there could be domain knowledge associated with relationships between features in the data. In datasets for autonomous vehicles, the data is multimodal since it can contain image data, sensor data, and behavior data. Jointly reducing the dimensionality of all the features is difficult since each of them can have unique characteristics, so finding reduced features from groups suggested by the domain (autonomous vehicles) that share a common characteristic is likely to be more effective (as just described in Sect. 10.1). Approaches that group features based on domain knowledge to reduce the dimensionality in the data have been shown to improve the performance of regression models [14, 16]. (A more detailed description of grouping-based PCA can be found in Sect. 10.1).

Example 10.5 For the malware classification problem [**MalC**] (described in Sects. 1.1 and 11.4), domain knowledge that certain features extracted from the software program are or should be related to others (e.g., they correspond to memory usage in the program) can help enhance the dimensionality reduction method through heuristics. Specifically, one can bundle features that are related and perform dimensionality reduction on different groups independently, instead of jointly performing dimensionality reduction on the full set of features. This grouping approach may lead to better reductions of the original feature set (as suggested in Sect. 10.1).   □

The malware dataset (defined in [1, 25] and summarized in Sect. 11.5) contains several well-known features for malware classification. Finding a reduced dimensional representation among the full set of a total of 350 features at once may not be ideal due to the inherent differences in the feature properties. A possible heuristic is to use knowledge about the domain (cybersecurity in this case) and group them into 6 different categories, for example as shown in Table 10.1. A dimensionality

Table 10.1 Grouping features using domain knowledge for reduction of the malware dataset

Group name	Brief description	# Features
Entropy	Measurements of disorder in the malware bytecode	204
Sections	Size of code sections	27
Assembler commands	Frequency of assembly commands used in the code	67
Registers	Frequency of registers used in the code	19
Data definitions	Proportion of data definition commands	13
Keywords	Manually chosen keywords in the metadata	20

reduction method (such as PCA in Sect. 4.1) or nonlinear geometric methods (as in Sect. 5.2) can now be performed on groups that have features with common properties, thus implicitly encoding domain knowledge in the dimensionality reduction method.

Specialized Methods

While the previous two approaches work on generic datasets, more sophisticated knowledge can help in dimensionality reduction for certain types of datasets, such as natural language/text processing. In this case, dimensionality reduction methods such as PCA will consider words as features and try to reduce dimensionality based on statistics seen in the data. However, the hidden semantic meaning (synonyms, antonyms) associated with words will always trump any correlations or directly observable statistics in the data. Specifically, for the task of comparing content in documents for semantic equivalence, if different words are used within the two documents, but the words have the same meaning, then the dimensionality reduction method should ideally generate a similar low-dimensional representation for both documents. A manual specification of word similarities has been shown to improve the performance of dimensionality reduction in text data [17]. That said, deep learning models have been able to extract rich semantic information from newly available massive datasets, without explicit domain knowledge. For instance, word embedding methods (such as Word2Vec [18]) exploit the fact that similar words tend to co-occur within the same context in documents in large-scale text corpora, and consequently, syntactic approaches have been able to achieve dimensionality reduction that factors in a good deal of semantic meaning of language. This is the key idea in latent semantic analysis. However, reducing dimensionality of a domain-specific vocabulary is a challenging problem [22]. For instance, in the specific domain of oil and natural gas considered in [22], embedding methods do not work as well as they use peculiar, rare words that are specific to the domain and so do not occur in general documents. The effectiveness of adding domain knowledge to deep-learning-based word embedding methods is an active area of research [10].

10.3   Heuristic Rules for Feature Selection, Extraction, and Number

There are many heuristic rules for feature selection and extraction. The rules are usually based on prior knowledge of the research questions or problems. Two examples are described below, but there are many more to explore.

Example 10.6 (Physics Rule/Knowledge of Predicting Velocity of An Object) A baseball batter naturally wants to predict the velocity v_t of a ball thrown by a pitcher, at certain time t (when it will hit the bat). She knows the ball's initial speed v_0 (pitcher's parameter) and the gravitational constant g. A common model can be built directly with three input variables

$$v_t = f(t, v_0, g),$$

and one response, the target velocity. However, by understanding the physical laws involved, one might just need to build a model based on one response and one input variable,

$$\pi_0 = g(\pi_1),$$

where $\pi_0 = v_t/v_0$, $\pi_1 = gt/v_0$. Although there are many different ways of formulating the key input variables [31], the main reason is that the underlying true model $v_t = v_0 + gt$ is equivalent to its simplest form $v_t/v_0 = 1 + gt/v_0$, that is, $\pi_0 = 1 + \pi_1$.   □

Example 10.7 (Agriculture Rule/Knowledge of Classifying Iris Flowers) It is well known that rich soil and more fertilizer can grow bigger flowers. In the Iris data, Table 11.7, only the sizes of the flowers are measured and reported: petal length, petal width, sepal width, and sepal length. However, to better classify the species, the shape of the flower is essential. Thus, the ratios of their lengths might contain more information than just lengths alone.   □

These examples illustrate that further analysis within the data's domain can provide insights into dimensionality reduction approaches, beyond the all-purpose methods suggested by data science.

10.4   About Explainability of Solutions

Explainability is an important aspect of data science solutions, particularly if the solutions are to be used in critical applications such as healthcare or national security, where decisions made have very high impact and/or cost. Recently, the European Commission has proposed regulatory measures for the use of data science solutions [12], and according to these regulations, *transparency* is an important aspect of such methods. It explicitly states that solutions should allow for human oversight, which implies a human should be able to understand the solution. This section characterizes the concept of explainability of a dimensionality reduction method or problem solution.

Example 10.8 For the classification problem for the Iris dataset [**IrisC**] (as described in Sects. 1.1 and 11.4), a nonexpert may inquire *WHY* a specific label is assigned by a certain classifier to a specimen of flower, especially if it falls on a borderline with another. A humanly understandable and rationalizable answer is expected, beyond "That's what the model says." One way to address the question is by ranking the relative importance of features that contributed to the label. That is, for a particular instance, the petal width may be more important to the classifier as compared to the sepal width. If it is possible to provide such a ranking of the features, then an expert in the domain (Iris flowers) can verify whether the explanation is indeed plausible, i.e., it makes sense across the labels he has seen in his long experience. This can help increase trust in classifiers that produces plausible explanations and at the same time can help in viewing results from classifiers with less plausible explanations (black boxes) with caution, even if such classifiers have high accuracy.   □

10.4.1   What Is Explainability?

Explainability has been studied in various disciplines such as psychology and cognitive sciences. In [21], the authors provide an overview of explainability from the perspectives of different disciplines. It turns out that what is viewed as explanation in data science is different from what has been considered as explanation in other fields such as psychology, cognitive science, and philosophy. In particular, data science explanations mainly focus on approximating complex decision-making functions in the model. Philosophy, on the other hand, has a long history in the study of explanations in terms of causality and epistemology. (A detailed primer on the influence of other sciences in explainability can be found in [19].) The term explanation (or explainability) as used in data science has received multiple interpretations, depending on the context in which it is used. In general, two important concepts underlying explainability of data science solutions include interpretability and transparency, according to [21]. Since the labels in a classifier are systematically produced by an algorithm behind it, ultimately the question is really about the model or algorithm behind it.

10.4.1.1   Outcome Explanations

Explainability is a property of a model that enables humans to interpret the decisions it makes. Thus, the explainability of a model can be characterized as its *degree of interpretability*, i.e., the extent to which the answers provided by the model are consistent across a body of knowledge and so can be rationalized by an expert mind. In general, explainability of a model involves two aspects. First, the model should provide or allow a ranking of the variables used based on their relative importance in arriving at a solution. Second, an external observer who has expertise in the domain should be able to corroborate the influence of the features influencing the answers with the ranking. Thus, if a model can relate its (complex) "black-box" decision-making process to a simpler known process within the realm of human common sense knowledge, then it is explainable.

Locally interpretable model-agnostic explanations (LIME) is a well-known general approach that explains predictions made by any complex classifier by ranking features based on their importance in the prediction. The main idea in LIME is to compute a local approximation for the classification decision boundary of a complex classifier. Specifically, given a data point whose prediction by the complex classifier needs to be explained, the boundary of the complex classifier near that data point is approximated with a simple linear decision boundary. The idea is that for the complex decision boundary, it is hard to rank the importance of features, but for the simple decision boundary, it is easier to do so. Thus, for any data point, LIME outputs a ranking of features that play the most important role in determining the response output by the complex classifier. (A full description of LIME can be found in [24].)

10.4.1.2 Model Explanations

Another approach is in terms of *global* explanations. Specifically, a model is *transparent* to a user if its operations can be explained for all inputs to the model. This is more general than the outcome/post-hoc explanations described in the previous subsection since it does not only explain how a model behaved for a specific instance but covers the functioning of the model over the entire population. That is, a model that is transparent to a user can explain how its decisions would change if certain inputs were altered, or what inputs need to be altered to change or obtain a certain classification. For example, if a model classifies an individual as having the flu based on some symptoms, an explanation could indicate the effect of changing a certain symptom on the flu prediction. This can help a user "debug" the data science model used to make the prediction. These explanations are also sometimes referred to as global explanations since they are not specific to a single instance. Borrowing from causal models [26], one approach to providing a global explanation is to establish a causal chain leading to an outcome. However, in general it may be possible to have multiple causes attributed to an outcome. Another potential direction is to use contrastive explanations, that is, explanations based on examples suggesting how the model would be different if the data was modified. Explainability of a data science model based on the degree of transparency is an active area of research. In general, explainability is hard to evaluate quantitatively since it is subjective. In practice, the standard approach to quantify explainability of a model is based on external knowledge. That is, another human is asked to give a "second opinion" that confirms or rejects the decisions or labels that the model outputs overall as an explainable solution to the problem.

10.4.2 Explainability in Dimensionality Reduction

The explainability of a dimensionality reduction method depends upon how it represents features in the lower-dimensional space. For example, consider the digit classification problem [**CharRC**] in Sect. 11.4. In this case, if the features directly correspond to pixels, then an interpretable explanation method can potentially measure the importance of these pixels in a particular classification model. A human can then easily verify this explanation. For example, if a classifier labels a digit image as the number 4, and the explainer points out that the pixels that are most useful in the classification are those near the crossing of the horizontal and vertical lines in the number 4, a human can verify that this explanation makes sense since that feature is indeed important in recognizing the digit 4. On the other hand, if the explainer points to random pixels, then the classification model will be less trustworthy. However, applying dimensionality reduction can reduce the explainability, since, for the features in the lower dimension, it may no longer be possible to map them to the original features.

In methods that use nonlinear combinations of features (summarized in Chap. 3), the features selected by the dimensionality reduction method become hard and sometimes impossible to interpret. Therefore, the existing explanation methods cannot give a valid explanation when using such dimensionality reduction methods. On the other hand, for each dimension output by PCA (described in Sect. 4.1), the original features and the coefficients for those features can be identified. Therefore, if PCA outputs for a lower dimension are considered important by the explainer, that explanation can be converted into an explanation on the original features. In general, to be able to explain using reduced data, the dimensionality reduction method needs to provide an inverse mapping that, given a dimension, returns the original features represented by that dimension. While recent work has been performed on interpreting nonlinear dimensionality reduction methods such as kernel PCA (described in Sect. 4.1) [15], this area remains under active research.

Example 10.9 Predictions made by a complex classifier such as a random forest [5] on the Iris dataset can be explained using the LIME explainer because LIME creates an interpretable linear model that approximates the classifier locally around the instance to be explained. It then ranks features in order of importance in the linear model. An example classification using LIME for random forests is illustrated in Table 10.2, according to [7]. Here, for specific instances of different classes in the [**IrisC**] problem, namely, Setosa, Versicolor, and Virginica, LIME specifies the relative importance of the features. The two most important features are shown in the table as the explanation for that instance. A domain expert can now verify this

Table 10.2 LIME explanations for some Iris dataset instances. PW is the petal width, PL is the petal length, SL is the sepal width, and SW is the sepal length. The top two features are shown as an explanation for an instance in each class, according to [7]

Class	PL	PW	SL	SW	Explanation
Setosa	1.2	0.2	4.4	2.9	(PW,PL)
Virginica	2.5	6.1	7.2	3.6	(PW,PL)
Versicolor	3.7	1	5.5	2.4	(PL,PW)

explanation to validate if indeed the petal width and length are the most important characteristics defining a kind of Iris flowers. If so, this would mean that the random forest classifier is indeed using the more relevant features for classification, thus increasing our trust in this model for the Iris flower classification problem.   □

10.5   Choosing Wisely

In this section, some rules that emerge from reflecting upon the methods presented in the previous chapters and sections are suggested as take-home messages on dimensionality reduction and solution models in data science. They are not universal rules that apply everywhere with good enough results, but just high-level meta-observations distilled from experience that have a number of specific applications and can be useful rules of thumb in unknown situations.

First of all, the *problem-solving methodology* discussed in Sect. 1.4.2 is extremely important because it may sound counterintuitive at times, especially to someone eager to tackle a problem directly and immediately. It is important to bear in mind that *haste makes waste*. If no problem/destination has been defined precisely (with details on input and expected answers, as well as the context for their eventual business application in the real world), no data can be appropriately selected and nothing will really happen purposefully. Every door will lead nowhere.

Second, *data selection, cleansing, and understanding* discussed in Sect. 1.2 require careful planning throughout, with the final destination in mind. One needs to be very mindful that *garbage in, garbage out* is a truism borne out of experience that will exact a dear toll if ignored.

Third, *dimensionality reduction* can be tackled in a number of ways, each with a particular criterion, and not every approach will be appropriate for any dataset, especially if optimal feature selection or extraction is desirable. Heuristically, one can only say that *generally* (a summary is in Table 10.3):

- *Grouping* features usually helps, especially if the available dataset is of a mixed, heterogeneous kind.
- *Statistical approaches* work well if the data is homogeneous and exhibits correlations across features.

Table 10.3 A comparison of DR methods along three major practical criteria (*column headers*) (A blank entry indicates a *No* answer)

DR method	Big data?	Explainable?	Parallelizable?
PCA	✓		
Regression		✓	✓
Shrinkage			✓
Entropy	✓	✓	✓
MDS			
ISOMAP			
t-**SNE**			
(RND)			
DNA	✓	✓	✓
Autoencoders	✓		✓
Neural nets	✓		✓

- *Feature selection* may work better on datasets that might have a small subset of important/relevant features.
- *Feature extraction* may work better on datasets that involve abstract relationships and semantics (e.g., text data and social media data).
- The *number of reduced features* may not be as important as the impact of the *structure of the data* in the quality of the reduction and/or the solution to a problem, as demonstrated by biomolecular approaches to DR and problem solving. On the other hand, it is easy to overdo it. For example, on average, solution models using only 6 features mostly performed poorly as compared to those using 12 features across in several problems in Chap. 6. Six (6) features might be too few for solving a complex problem, so reducing dimensionality too much may hurt for abiotic data, while it is quite possible for DNA sequences, as demonstrated by the impressive reductions with nxh chips in Chap. 7.
- *Computational running times* may be determinant of the success of an approach to DR or a solution, particularly for big data. PCA is very effective because it can be done with matrix methods for which libraries have been developed and optimized, but only to a point since their optimization is done globally and all the features need to be considered at once. On the other hand, entropy-based approaches in Chap. 6 offer the additional advantage of being computationally efficient because they are easily parallelizable on any platform (e.g., using MapReduce/Hadoop). They can select features in the order of minutes (single entropies), under 2 hours (paired entropies) and under 6 hours (iterated entropies); for the nxh basis mentioned in Chap. 7, feature extraction on an HPC (using a single node) also takes in the order of minutes for all datasets. Further, these times essentially scale linearly with the size of the data, except for iterated entropy. This makes the process of feature extraction feasible, even for huge datasets. However, the systematic good performance of iterated entropy methods comes at a dear price because the running times of extracting the features increase exponentially with the dimensionality of the data.

Table 10.4 A comparison of DR solution methods along three major practical criteria (*column headers*) (A blank entry indicates a *No* answer)

DR method	Big data?	Explainable?	Parallelizable?
Regression	✓	✓	
Decision trees		✓	✓
Bayesian learning	✓	✓	✓
kNN		✓	✓
Ensemble methods	✓		✓
SVMs	✓		✓
Neural nets	✓		✓
k-Means	✓	✓	✓
DNA-based	✓	✓	✓

- *Explainability* of the solutions to human customers may be extremely important for some applications. For example, machine learning methods based on deep learning and convolutional neural nets may be hard to rationalize, while information-theoretic methods and biomolecular computing methods have a domain foundation behind them that makes them easier to justify. Thus, the applications produced from these solutions may meet more acceptability than those produced by other methods since they can be easily rationalized.

Fourth, for *solution models*, data science problems can be tackled in a number of ways, each with particular strengths and blind spots, and not every model will be appropriate for every problem. Table 10.4 only suggests some *general* heuristics that may perfectly well prove ill-suited for a given problem. Since the solution models are data dependent, it is hard to give exact rules for choosing one without experimentally evaluating them using standard metrics since every method is data dependent. However, below are some typical heuristics that could help us choose solution models:

- GLM
  Generalized linear models are most useful when the response variable is a discrete categorical variable. In particular, GLM includes the popular logistic regression model as an important special case.
- DTs
  Decision trees are arguably the most interpretable machine learning methods. However, decision tree optimization is computationally hard. Therefore, they are less suited for big data applications. New approaches have been explored to parallelize decision trees to improve scalability.
- Bayesian classifiers
  Naive Bayes classifiers are useful when each feature is independent of other features since that is the underlying assumption. However, even when the assumption is not strictly met (e.g., text processing), Naive Bayes is known to obtain accurate results in practice. It is highly scalable and therefore is well suited for big data applications.

- **kNN**
  k-Nearest neighbors typically work well for lower-dimensional data (i.e., when we have a small number of features). In higher-dimensional data, computing neighbors is harder (due to the curse of dimensionality, as described in the next section). Also, k-Nearest Neighbors scale poorly to big datasets since computing neighborhoods over large datasets is a hard computational problem.
- Ensemble methods
  Typically, bagging is used when the base classifier overfits, i.e., one obtains very high training accuracy but poor test accuracy. Using bagging in this case helps us reduce variance in the classifier and reduce overfitting. Further, bagging is more helpful with unstable learning algorithms [6], where the classifier changes significantly with a slight change in the training data. On the other hand, boosting is typically more helpful when the classifier is underfitting, i.e., it has poor accuracy on the training data. Boosting will combine several such weak classifiers together and improve their performance.
- k-Means
  k-Means clustering works best when there is a good separation of the data, and the clusters have a globular shape. However, it is not always guaranteed to find the optimal clusters. Further, if there are outliers in the data, k-Means tends to return poor clusters. However, it is the method of choice for large datasets since it is highly scalable.
- SVMs
  Support vector machines work well in problems that have a large number of features (e.g., text processing). In such cases, a linear SVM can find a good decision boundary since it maximizes the margins, i.e., distance from the decision boundary to support vector instances in either class in a binary classification problem. In higher-dimensional data, the use of kernels may sometimes overfit since dimensionality implicitly increases using a kernel. When the dataset is imbalanced, SVMs typically perform poorly since they learn classifiers that are biased toward the majority class.
- Neural nets
  Neural nets are exceptionally suited to provide good enough solutions to just about any problem one throws at them, as demonstrated by nearly 50 years of practical experience now. As a consequence of the NFL theorem (described in Sect. 10.7 below), no model can provide optimal solutions most of the time. They are therefore a versatile tool one needs to have handy in any toolbox.

10.6   About the Curse of Dimensionality

The curse of dimensionality is a term coined by Richard Bellman [3] in 1957, who used it to refer to the fact that using grid search to solve optimization problems with many variables (high dimensions) is very inefficient and practically infeasible. Subsequently, the term has been used in a variety of contexts. Though the details of each context differ, the underlying fundamental reason is the same as was introduced in Sect. 3.3, and it can be intuitively described by the aphorism—"there is too much space in high dimensions." We elucidate this using some simple calculations and then mention how this affects tasks in data analysis.

It is well known the area of a circle is proportional to the square of its radius and the volume of a three-dimensional ball is proportional to the cube of its radius. The same phenomena hold in higher dimensions. Let $B_d(c, r)$ denote the ball of radius r centered at a point c, and let $v_d(r)$ denote its volume, i.e., $v_d(r) = v_d(B_d(c, r))$, where $v_d(\cdot)$ denotes the d-dimensional volume. It is well known that

$$v_d(r) = \frac{\pi^{d/2} r^d}{\Gamma(d/2 + 1)},$$

where $\Gamma(\cdot)$ is the Gamma function [2]. The following example will illustrate one of the computational consequences of the above formula.

Example 10.10 Grid search is an important method to optimize hyperparameters in deep learning (Sect. 2.2.4). An artificial but representative setting is the following— this can arise for example if we are using hyperparameters with some regularization. Suppose we want to minimize a certain function $F(x_1, x_2, \ldots, x_d)$, where the x_i are the hyperparameters, and a regularization condition is that they lie within the ball of radius r, i.e., $\sum_{i=1}^{d} x_i^2 \leq r^2$. If the function $F(x_1, \ldots, x_d)$ is complicated as for example the loss function for a deep neural network, this may not be feasible analytically. One way to accomplish this practically is to sample from the ball of radius r so that for each point of the ball, we have a sample point within a distance of δr where δ is a small number, say 0.1. Such a set of points is known as a *net*. Then, assuming the function $F(x_1, \ldots, x_d)$ is continuous (or varies slowly), it would make sense to compute its minimum over a net and expect the computed minimum to be not too far from the actual minimum. This would require us to compute $F(x_1, \ldots, x_d)$ for each point of the net. As such, we would like the size of such a net to be as small as possible. How large is the smallest possible net? In $d = 1$ dimensions, it can be verified that there is a net of size $2\lceil 1/\delta \rceil + 1$, for example by taking the points $0, \delta, 2\delta, \ldots, \min(1, \lceil 1/\delta \rceil)$. While it is not so easy to explicitly enumerate such a net in higher dimensions, we remark that actually constructing such a net is not too hard for example by finding all points on a sufficiently fine grid inside the ball. However, there are constraints.

If S is any grid as described above, then $|S| \geq \left(\frac{1}{\delta}\right)^d$.

The following proof, which can be skipped if the reader so chooses, is to illustrate the role the formula for volume mentioned above plays here.

Proof Since every point of the ball $B_d(c, r)$ is within δr of a net point, we have that

$$B_d(c, r) \subseteq \bigcup_{p \in S} B_d(p, \delta r).$$

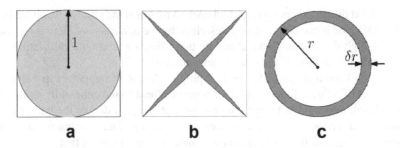

Fig. 10.1 (a) The unit ball is a good approximation to the unit cube in 2 dimensions. (b) A depiction of a projection to 2 dimensions of a ball in very high dimensions—it looks spiky like a porcupine. (c) A shell of width δr in a ball of radius r in $\Re^d$. For $\delta = 0.1$ and $d = 100$, the shell has 99.9% of the ball volume

This implies, by elementary properties of the volume,

$$v_d(B_d(c, r)) = v_d(r) \leq \sum_{p \in S} v_d(B_d(p, \delta r)) = |S| v_d(\delta r).$$

Putting in the formula above for $v_d(\cdot)$ and simplifying, we get

$$|S| \geq \left(\frac{1}{\delta}\right)^d,$$

as claimed.

To see what this means, consider a situation where we are working in $d = 100$ dimensions and $\delta = 0.1$. The size of such a net is then at least 10^{100}. Storing or iterating over a net, this large is computationally infeasible, and thus such a grid search method is practically infeasible in this range. We need to look for alternative methods or else satisfy ourselves with a smaller accuracy and thereby a smaller net.

The geometry of high dimensions can also be very counterintuitive. For example, consider the cube $[-1, 1]^d$. In $d = 2$ dimensions, the ball of radius 1 with center at the origin is a good approximation to the cube, see Fig. 10.1a. In fact, we cannot pack more than one such ball into the cube. However, in higher dimensions, such a ball is a very bad approximation to the cube. Indeed, the volume of the cube is 2^d, and the volume of the ball is $\pi^{d/2}/\Gamma(d/2 + 1)$. One can show by using simple asymptotic approximations of the Gamma function that the volume of the ball of radius 1 is

$$v_d(1) \sim \frac{(\pi e)^{d/2}}{\sqrt{2\pi}(d/2)^{d/2+1}},$$

which is already miniscule even for $d = 20$. This is quite contrary to the intuition, particularly, since despite this fact that the volume of the unit ball is extremely small compared to the cube, there is only one unit ball that can fit inside the cube! Indeed, if the center of such a ball is anything other than the origin, part of such a ball will be outside the cube. Thus our imagination of the cube being a well-rounded body is sometimes described as an inaccurate picture. Probably, a more accurate description is that the cube is "spiky" in high dimensions, and its center (that can fit the unit ball) is a very small region, while most of the volume is carried by the spikes—in other words, the cube looks like a porcupine. Figure 10.1b shows a representation of a 2-dimensional projection of such a cube in very high dimensions. This spikiness can be attributed to the fact that measure tends to concentrate near the boundary, also mentioned in Sect. 3.3. We show a small calculation demonstrating this to develop some intuition about the result.

Example 10.11 Imagine a ball of radius r centered at the origin. Consider a shell inside the ball of width δr, see Fig. 10.1c. What is the measure inside this shell compared to the measure inside the ball? To compute this, we find its volume and divide it by the volume of the ball. This is precisely

$$\frac{\mathsf{v}_d(r) - \mathsf{v}_d((1 - \delta)r)}{\mathsf{v}_d(r)}.$$

Using the formula for $\mathsf{v}_d(\cdot) = \frac{\pi^{d/2}r^d}{\Gamma(d/2+1)}$, this readily simplifies to $1 - (1 - \delta)^d \geq 1 - e^{-\delta d}$, which holds because of the inequality $1 - x \leq e^{-x}$. Suppose $d = 100$ and $\delta = 0.1$, i.e., we are just about 10% into the ball from its boundary to its center. In this case, our estimate gives us that the shell contains at least $1 - e^{-10} > 0.999$; i.e., 99.9% of the measure of the entire ball is within the shell.

A consequence of this concentration of measure phenomena is its implication for nearest neighbor search where the curse of dimensionality has serious consequences that cannot be circumvented by any algorithmic technique. As was shown by Beyer et. al [4], if one assumes that data points are selected independently from certain distributions, and the query point is selected independently as well, then for any fixed $\varepsilon > 0$ as the dimension increases, the distance of the query point to all the data points is within $(1 + \varepsilon)$ of its distance to the nearest neighbor, with high probability. Thus all points appear roughly equally close, rendering the entire concept of nearest neighbors useless in some sense. The distributions mentioned above are not too restrictive, and in particular, they include but are not limited to product distributions such as the uniform distribution on a cube. On the other hand, this result does not hold for example when the dimensions are highly correlated, and therefore, under the manifold hypothesis mentioned before (Sect. 5.1), one can still hope that proximity queries are useful.

10.7 About the No-Free-Lunch Theorem (NFL)

In pursuit of ever more general procedures to perform any job, e.g., problem solving in any science (data science in particular) *in a most general and optimal way*, humans are always faced with Shakespeare's Macbeth's dictum "[...] the desire is boundless, and the act a slave to the limit." Every science ultimately unearths limitations inherent in the phenomena it concerns. The second law of thermodynamics, relativistic limits on the speed of light in physics, and Godel's incompleteness theorems in logic are but examples. So it is not surprising that machine learning and data science have also discovered inherent limitations in how far the power of their methods can go. The No-Free-Lunch (NFL) theorem is one prime example. According to the seminal paper by Wolpert and Macready [8]:

> We have dubbed the associated results *NFL theorems* because they demonstrate that if an algorithm performs well on a certain class of problems, then it necessarily pays for that with degraded performance on the set of all remaining problems.

So, what exactly is the "No Free Lunch Theorem" (NFL)? It is a practical (backed up by a theoretical) finding that essentially says that all optimization algorithms perform equally well when their performance is averaged over all possible objective functions. In the field of machine learning, NFL implies that all optimization algorithms perform equally well when their performance is averaged across all possible problems. Consequently, it suggests that there cannot exist a single best machine learning algorithm for solving problems such as classification, regression, clustering, or prediction. (Further elaboration on NFLs can be found at [8].)

In machine learning problem and data science, it is common to use an algorithm to figure out the relationship or function mapping between the features X and a target Y. To build a model for the purpose of making predictions or solving a problem, every machine learning algorithm makes its assumptions about the data. Obviously, the "goodness" of the model depends on how well the underlying patterns in the data fit the assumptions. Therefore, any particular learning method may not perform well if the assumptions failed. One cannot have the "best" predictions, even with perfect knowledge of the underlying data. Hence, a particular method most likely to be best for one problem or even dataset is not necessarily going to work well for other data with different characteristics, let alone other problems, even if apparently similar. Therefore, it is the data scientist's job to explore and understand data carefully in order to develop the right model and various models to find good enough solutions. It is important to know about the assumptions made by each algorithm so that one can choose the right method to use. In summary, for a data scientist, the aim should not be to find a single "best" method to solve all problems, but to find methods that produce good enough solutions for a few problem at hand.

A famous result characterizing the difficulty in machine learning was proposed by Wolpert [30] and Schaeffer [27]. Following this, a series of other formalizations of the result for other tasks such as search and optimization [30] were developed. These results are collectively called the *No-Free-Lunch* (NFL) theorems. Specifically, the NFL theorems outline the implicit tradeoffs associated with learning a model based on data. In particular, the conservation law of generalization performance [27] proves that *no learning bias can outperform any other bias over the space of all possible learning tasks*. Here, a learning bias is an assumption made by the learning algorithm. For example, in decision trees (described in Sect. 2.1), an implicit bias is to prefer shallow-depth trees; in PCA (described in Sect. 4.1), the bias is to learn linear combination of features to represent the original features. The NFL result implies that no matter what assumptions one makes, there will be a subset of problems for which these assumptions will fail to yield a good enough model.

As a concrete example of NFL, several learning algorithms refer to what is called as *Occam's razor* to bias the learning such that simpler models are specified over more complex ones [29]. However, there follows from NFL that a learning bias such as Occam's razor, though valuable in some tasks, can only be useful for some subset of all possible learning tasks. In other words, there is no universal learning bias that provably works for any learning task. However, if one argues that real-world learning tasks are a subset of all possible learning tasks where such learning biases work well, then even with the somewhat negative result of the NFL theorems, practically, it is perfectly reasonable to use these biases within learning and dimensionality reduction algorithms. In fact, in a related work [23], the argument is made that real-world learning tasks are not uniformly distributed across the possible learning tasks. A reasonably sound argument for why this is the case is based on the fact that real-world learning tasks are a product of human knowledge. For instance, to humans, in the type of problems that machine learning is useful for, the independent variables are likely to be related to the dependent variable in ways that can be inductively inferred by classifiers. Thus, despite negative results from NFL, it is still possible to have an optimistic view of assumptions made in learning and dimensionality reduction methods in general, if one is thoughtful enough to make the right assumptions.

References

1. Ahmadi, M., Ulyanov, D., Semenov, S., Trofimov, M., & Giacinto, G. (2016). Novel feature extraction, selection and fusion for effective malware family classification. In *Proceedings of the Sixth ACM Conference on Data and Application Security and Privacy* (pp. 183–194).
2. Artin, E. (1964). *The gamma function*. New York: Holt Rinehart and Winston.
3. Bellman, R. E. (2003). *Dynamic programming*. New York, USA: Dover Publications.
4. Beyer, K., Goldstein, J., Ramakrishnan, R., & Shaft, U. (1999). When is "nearest neighbor" meaningful? In *International Conference on Database Theory (ICDT '99)* (pp. 217–235). Berlin: Springer.
5. Breiman, L. (2001). Random forests. *Machine Learning, 45*(1), 5–32.
6. Breiman, L. (1996). Bagging predictors. *Machine Learning, 24*(2), 123–140.
7. Burkart, N., & Huber, M. F. (2021). A survey on the explainability of supervised machine learning. *Journal of Artificial Intelligence Research* **70**, 245–317.
8. David, H. W., & William, G. M. (1997). No free lunch theorems for optimization. *IEEE Transactions on Evolutionary Computation, 1*(1), 67–82 (1997).

9. Davidson, I. (2009). Knowledge driven dimension reduction for clustering. In C. Boutilier (Ed.), *IJCAI 2009, Proceedings of the 21st International Joint Conference on Artificial Intelligence, Pasadena, California, July 11–17, 2009* (pp. 1034–1039).

10. Deng, C., Ji, X., Rainey, C., Zhang, J., & Lu, W. (2020). Integrating machine learning with human knowledge. *iScience, 23*(11), 101656.

11. Douglas, B. L. (1995) CYC: A large-scale investment in knowledge infrastructure. *Communications of the ACM, 38*(11), 33–38.

12. European Commission (2020). *White paper on artificial intelligence: A European approach to excellence and trust.* White Paper COM (2020) 65 Final, European Commission.

13. Fellbaum, C. (1998). *WordNet: An electronic lexical database.* Bradford Books.

14. Guo, Z., Li, L., Lu, W., & Li, B. (2015). Groupwise dimension reduction via envelope method. *Journal of the American Statistical Association, 110*(512), 1515–1527.

15. Hosseini, B., & Hammer, B. (2019). Interpretable discriminative dimensionality reduction and feature selection on the manifold. In *ECML/PKDD (1)*. Lecture Notes in Computer Science (Vol. 11906, pp. 310–326). Springer.

16. Li, L., Li, B., & Zhu, L. (2010). Groupwise dimension reduction. *Journal of the American Statistical Association, 105*(491), 1188–1201.

17. Mao, Y., Balasubramanian, K., & Lebanon, G. (2010). Dimensionality reduction for text using domain knowledge. In *COLING (Posters)* (pp. 801–809). Chinese Information Processing Society of China.

18. Mikolov, T., Sutskever, I., Chen, K., Corrado, G. S., & Dean, J. (2013). Distributed representations of words and phrases and their compositionality. In *Neural Information Processing Systems* (pp. 3111–3119).

19. Miller, T. (2019). Explanation in artificial intelligence: Insights from the social sciences. *Artificial Intelligence, 267*, 1–38.

20. Mitchell, T., Cohen, W., Hruschka, E., Talukdar, P., Yang, B., Betteridge, J., Carlson, A., Dalvi, B., Gardner, M., Kisiel, B. Krishnamurthy, J. Lao, N., Mazaitis, K., Mohamed, T., Nakashole, N., Platanios, E., Ritter, A., Samadi, M., Settles, B., . . . , Welling, J. (2018). Never-ending learning. *Communications of the ACM, 61*(5), 103–115.

21. Mittelstadt, B., Russell, C., & Wachter, S. (2019). Explaining explanations in AI. In *Proceedings of the Conference on Fairness, Accountability, and Transparency* (pp. 279–288).

22. Nooralahzadeh, F., Øvrelid, L., & Lønning, J. T. (2018). Evaluation of domain-specific word embeddings using knowledge resources. In *Proceedings of the Eleventh International Conference on Language Resources and Evaluation (LREC 2018)*.

23. Rao, B. R., Gordon, D. F., & Spears, W. M. (1995). For every generalization action, is there really an equal and opposite reaction? In A. Prieditis & S. J. Russell (Eds.), *International Conference on Machine Learning* (pp. 471–479). Morgan Kaufmann.

24. Ribeiro, T. M., Singh, S., & Guestrin, C. (2016). "Why Should I Trust You?": Explaining the predictions of any classifier. In *Knowledge discovery and data mining* (pp. 1135–1144) ACM.

25. Ronen, R., Radu, M., Feuerstein, C., Yom-Tov, E., & Ahmadi, M. (2018). Microsoft malware classification challenge. *CoRR, abs/1802.10135*.

26. Pearl, J. (2009). *Causality* (2nd ed.). Cambridge: Cambridge University Press.

27. Schaffer, C. (1994). A conservation law for generalization performance. In *International Conference on Machine Learning, ICML* (pp. 259–265). Morgan Kaufmann.

28. Swartout, W. R. (1985). Rule-based expert systems: The mycin experiments of the stanford heuristic programming project. *Artificial Intelligence, 26*(3), 364–366.

29. Wolpert, D. H. (1990). The relationship between Occam's razor and convergent guessing. *Complex Syst, 4*(3), 319–368.

30. Wolpert, D. H. (1996). The lack of a priori distinctions between learning algorithms. *Neural Computation, 8*(7), 1341–1390.

31. Yang, C. C., & Lin, D. K. J. (2021), A note on selection of basis quantities for dimensional analysis. *Quality Engineering, 33*(2), 240–251 (2021).

Chapter 11
Appendices

Max Garzon ⓘ **, Lih-Yuan Deng** ⓘ **, Nirman Kumar** ⓘ **, Deepak Venugopal,**
Kalidas Jana, and Ching-Chi Yang ⓘ

Abstract This chapter presents a summary review of prerequisite concepts from statistics, mathematics and computer science, although readers are expected to have a nodding familiarity with most of them. It also provides some background on a number of computational problems and data sets used in the book or particularly useful for data science and dimensionality reduction; as well as a review of computing environments and platforms that could be used as a playground to run and test the methods and solutions described in this book. The aim is to provide a refresher of what they are and point to sources in the literature where they could be studied in more detail, if needed.

Basic concepts such as "element," "set," "subset," "power set" of a given universe or populations of objects Ω and operations among them (such as "union," "intersection," "complementation," and such) and "functions" among them will be assumed to be familiar to the reader so as to require no definition beyond their ordinary intuitive meaning from elementary high school or college mathematics. For example, the *indicator* function χ of a subset $E \subset \Omega$ is a function with domain Ω into the real numbers **R**, given by $\chi_E(s) = 1$ if $s \in E$ and $\chi_E(s) = 0$ in the complement of E.

M. Garzon (✉) · N. Kumar · D. Venugopal
Computer Science, The University of Memphis, Memphis, TN, USA
e-mail: mgarzon@memphis.edu; nkumar8@memphis.edu; dvngopal@memphis.edu

L.-Y. Deng · C.-C. Yang
Mathematical Sciences, The University of Memphis, Memphis, TN, USA
e-mail: lihdeng@memphis.edu; cyang3@memphis.edu

K. Jana
Fogelman College of Business, Memphis, TN, USA
e-mail: kjana@memphis.edu

11.1 Statistics and Probability Background

The subject matter of *statistics* is *understanding and processing data at the level of aggregates and populations through observations of their individual objects*. The subject matter of *probability* is *random phenomena and their inherent uncertainty*. A cornerstone concept for both is that of a random *experiment*, i.e., any process in the world that can be repeated any number of times and produces one of a certain number of observations (or *outcomes*) every time, such as tossing a coin, rolling a die, rolling a pair of dice, the total rainfall amounts at home on a given day, the life length of the only light bulb in the kitchen, or the time between consecutive emissions of atomic particles from this sample of a radioactive material. The set of single outcomes (called *simple* or *elementary*) *events* of such an experiment is called the *sample space* Ω of the experiment. For example, for the toss of the same coin twice, the sample space is (Table 11.1)

$$\Omega = \{(H, H), (H, T), (T, H), (T, T)\}.$$

Probabilistic analysis proceeds by assigning a measure of *likelihood* or *uncertainty* to these elementary events and their combination into more complicated compound events (all referred to as just *events*), like parity ($\{2, 4, 6\}$) in the case of the roll of a die. These measures are usually assigned by frequency of occurrence in a large number of repetitions of the experiment, but they can be arbitrarily assigned as long certain rules of common sense are respected. Such a structure is called a probability space.

A *probability space* consists of a sample space Ω and a function P on Ω that assigns a real value $P(s)$ to every simple event $s \in \Omega$ and by extension, a probability value to composite events E according to

$$P(E) = \sum_{s \in E} P(s),$$

so that $P(\varnothing) = 0$ and $P(\Omega) = 1$.

In particular, if an event $A = A_1 \cup \cdots \cup A_k$ can be partitioned into a *disjoint* set of parts A_i, then $P(A) = P(A_1 \cup \cdots \cup A_k) = \sum_i P(A_i)$, whereas only $P(A) = P(A_1 \cup \cdots \cup A_k) \leq \sum_i P(A_i)$ is true if the subsets have overlaps. If Ω is finite, this procedure assigns a probability to every subset of Ω. For infinite sets in the continuum, the assignment is more involved (with sums of an infinite number of

Table 11.1 Sample space and possible values of a random variable X that counts heads in the random experiment of tossing a coin twice

Simple event s	First toss	Second toss	$X(s)$
(H, H)	1	1	2
(H, T)	1	0	1
(T, H)	0	1	1
(T, T)	0	0	0

terms), but P usually boils down to using measures such as length (1D intervals), areas (2D regions), volumes (in 3D spaces) and perhaps hypervolumes (in higher dimensional spaces), and follow the same basic rules.

Probability spaces can be classified as *discrete* or *continuous* depending on whether the sample space is a finite or countable set, or whether it is larger in cardinality, say the continuum of a line segment in 1D Euclidean space, where the underlying random experiment could be taking a random point from the unit interval $[0, 1]$.

In a probability spaces, the *conditional probability* of an event B given another event A is

$$P(B \mid A) = P(B \cap A)/P(A)$$

Two events A, B in a probability space are *stochastically independent* if the outcome of the random experiment being in A does not affect the probability of it being in B, i.e., if $P(B \mid A) = P(B)$, or equivalently, $P(A \cap B) = P(A)P(B)$.

A probability function P is a special case of a more general concept of random variable as observations made on the outcomes of a random experiment.

A *random variable* (RV) is a real-valued function $X : \Omega \rightarrow \mathbf{R}$ defined over the probability space (Ω, P), i.e., for each $s \in \Omega$, $X(s)$ is a (properly defined) real value. A (discrete) *stochastic process* is a sequence of random variables $\{X_t\}_{t \in \mathbf{N}}$ defined on a common probability space. A *sample* of a probability space (Ω, P) are the values of a run of the stochastic process $(X_{t \leq n})$ for some finite value $n \in \mathbf{N}$ selected according to the same probability model P (so the variables are *independent and identically distributed (i.i.d.)* .)

Example 11.1 The number of heads X when tossing two coins (all possible values of X are 0, 1, or 2), and a value between 1 and 6 for the roll of a die are properly defined RVs, but the name or age of the person who tossed the coin(s) is not since the coin(s) tosses do not determine it.   □

RVs can also be classified as *discrete* or *continuous* depending on whether the set of values taken on is a countable set (like a subset of the natural numbers $\mathbf{N}$, as with counting heads on coin tosses) or not (like the continuum of values in the real numbers $\mathbf{R}$, as with some measurement of a continuous quantity, like length, weight or time.)

Any random variable X defines certain important events for given values a of one observation. For example, the event denoted $[X = a]$, $(X = a)$, or simply $X = a$, consists of the elementary events that produce a specific observation with value a. The probabilities of these events define an important *distribution* associated with X.

The *probability density function* (pdf), also called *mass function* (pmf), of X is the function f_X (or just f if X is clear from the context) given by

$$f(x) = P(X = x), \quad x \in \mathbf{R}.$$

For continuous random variables that can take on infinitely many values in $\mathbf{R}$ corresponding to points on an interval, the values could be nonzero despite the fact that there is an infinite number of them and all their probabilities must still add up to 1. Therefore, this definition still makes sense for continuous RVs X as well, but the calculations get more involved and it is not really required for the purpose of this book.

Tables 11.2 and 11.3 show characteristic properties and facts about commonly used discrete and continuous probability distributions, respectively.

In statistics and data science, understanding a RV may be difficult because of the various values it may take in a variety of situations (say, the salaries of American citizens today.) A human mind makes better sense of a few values, so statisticians are challenged with the data science task of making RVs accessible to human minds. For example, what *single value to choose as most representative* of the values taken on by the RV? How spread are the values of the variable across the population Ω?

The *expected value* (or *mean*) $\mu = E(X)$ of a RV X is the weighted sum of the values of X using as weights the probabilities P_k of the events $[X = x_k]$, i.e., for discrete X,

$$E(X) = \sum_{x \in \mathbf{R}} x f(x) = \sum_k x_k P(X = x_k) = \sum_k P_k x_k,$$

where the x_ks range over all the values taken on by X; and for continuous X,

$$E(X) = \int_{x \in \mathbf{R}} x f(x) \, dx$$

Table 11.2 Common discrete RVs, probability distributions and their characterizing parameters (pdf $f(x)$, mean μ and variance σ^2)

Distribution $X \sim$ parameters Range of X	pdf	Mean	σ^2
Standard uniform $X \sim U(1, \ldots, m)$ $\{1, \cdots, m\}$	$\frac{1}{m}$	$(m+1)/2$	$\frac{m^2-1}{12}$
General uniform $X \sim U(A, \ldots, B)$ $\{A, A+1, \cdots, B\}$	$\frac{1}{B-A+1}$	$(A+B)/2$	$\frac{(B-A+1)^2-1}{12}$
Bernoulli $X \sim Ber(P)$ $\{0, 1\}$	$P^x(1-P)^{n-x}$	P	$P(1-P)$
Binomial $X \sim B(n, P)$ $\{0, 1, 2, \cdots, n\}$	$\binom{n}{x} P^x(1-P)^{n-x}$	nP	$nP(1-P)$
Poisson $X \sim Poisson(\lambda)$ $\mathbf{N} = \{0, 1, \cdots\}$	$\frac{\lambda^x e^{-\lambda}}{x!}$	λ	λ
Geometric $X \sim Geo(P)$ $\mathbf{N} = \{0, 1, \cdots\}$	$P(1-P)^x$	$\frac{1-P}{P}$	$\frac{1-P}{P^2}$
Negative binomial $X \sim NB(r, P)$ $\mathbf{N} = \{0, 1, \cdots\}$	$\binom{x+r-1}{r-1} P^r(1-P)^x$	$\frac{r(1-P)}{P}$	$\frac{r(1-P)}{P^2}$
Hypergeometric $X \sim HG(n, N, S)$ $\{L, \cdots, U\}$	$\frac{\binom{S}{x}\binom{N-S}{n-x}}{\binom{N}{n}}$ where $L = \max(0, n+S-N)$	$n\frac{S}{N}$ and	$n\frac{S}{N}\frac{(N-S)}{N}\frac{N}{(N-1)}$ $U = \min(n, S)$

Table 11.3 Common continuous RVs, probability distributions and their characterizing parameters (pdf $f(x)$, mean μ and Variance σ^2)

Distribution $X \sim$ parameters	pdf	Mean	σ^2
Range of X			
Standard uniform $X \sim U([0, 1])$	$f(x) = 1$	1	$\frac{1}{12}$
Interval $[0, 1]$			
General uniform $X \sim U([A, B])$	$f(x) = \frac{1}{B-A}$	1	$\frac{B-A}{12}$
Interval $[A, B]$			
Standard normal $X \sim N(0, 1)$	$f(x) = \frac{1}{\sqrt{2\pi}} e^{-x^2/2}$	0	1
$\mathbf{R} = (-\infty, +\infty)$			
General normal $X \sim N(\sigma, \mu)$	$f(x) = \frac{1}{\sqrt{2\pi}\sigma} e^{-(x-\mu)^2/(2\sigma^2)}$	μ	σ^2
$\mathbf{R} = (-\infty, +\infty)$			
Standard logistic $X \sim Logis(0, 1)$	$f(x) = \frac{e^{-x}}{(1+e^{-x})^2}$	0	$\frac{\pi^2}{3}$
$\mathbf{R} = (-\infty, +\infty)$			
General logistic $X \sim Logis(\mu, \beta)$	$f(x) = \frac{1}{\beta} \frac{e^{-(x-\mu)/\beta}}{(1+e^{-(x-\mu)/\beta})^2}$	μ	$\frac{\pi^2}{3\beta^2}$
$\mathbf{R} = (-\infty, +\infty)$			
Standard exponential: $X \sim Exp(1)$	$f(x) = e^{-x}$	1	1
$[0, +\infty)$			
General exponential: $X \sim Exp(\lambda)$	$f(x) = \lambda e^{-\lambda x}$	$\frac{1}{\lambda}$	$\frac{1}{\lambda^2}$
$[0, +\infty)$			
Double exponential: $X \sim DExp(\lambda)$	$f(x) = \frac{\lambda}{2} e^{-\lambda |x|}$	0	$\frac{2}{\lambda^2}$
$\mathbf{R} = (-\infty, +\infty)$			
Gamma: $X \sim Gamma(\alpha, \beta)$	$f(x) = \frac{1}{\Gamma(\alpha)\beta^\alpha} x^{\alpha-1} e^{-x/\beta}$	$\alpha\beta$	$\alpha\beta^2$
$[0, +\infty)$			
Beta: $X \sim Beta(\alpha, \beta)$	$f(x) = \frac{\Gamma(\alpha+\beta)}{\Gamma(\alpha)\Gamma(\beta)} x^{\alpha-1}(1-x)^{\beta-1}$	$\alpha\beta$	$\alpha\beta^2$
$[0, 1]$			
$\chi^2 : X \sim \chi^2(v/2, 2)$	$f(x) = \frac{1}{\Gamma(v/2)2^{v/2}} x^{v/2-1} e^{-x/2}$	v	$2v$
$[0, +\infty)$			
Student t: $X \sim t(v/2, 2)$	$f(x) = \frac{\Gamma v+1}{\Gamma v\sqrt{v\pi}} \left(1 + \frac{x^2}{v}\right)^{-\frac{v+1}{2}}$	$0\ (v > 1)$	$\frac{v}{v-2}$
$\mathbf{R} = (-\infty, +\infty)$			$(v > 2)$

because the infinitely many terms in the sum force an integral.
The *variance* of X is

$$Var(X) = E(X - E(X)) = \sum_{x \in \mathbf{R}} (x - \mu)^2 f(x) \quad \text{or} \quad \int_{x \in \mathbf{R}} (x - E(X))^2 f(x) dx$$

i.e., the mean of the RV $Y = X - E(X)$. The *standard deviation* (*std*) of X is

$$\sigma = \sqrt{Var(X)} \ .$$

These parameters $\mu = E(X)$ and $\sigma(X)$ provide useful statistics for humans to get an intuitive understanding of a RV X. Besides having single numbers summarizing a dataset, they are useful in many other ways. For example, the features in a given dataset $\mathbf{X}$ may be scaled different for different features and common scales may be desirable. One can *center* the data by subtracting from each feature X_i its mean μ_i, so that the mean of the features are then all common and equal to 0. Further, one can also rescale the centered values by their standard deviation σ, i.e., replace X_is by $(X_i - \mu_i)/\sigma_i$. This is the *standard normalization* of the new features to mean 0 and std 1 and are called the *Z-scores*. They exhibit essentially the same statistical properties of the original data $\mathbf{X}$, but may facilitate other methods (e.g., Principal Component Analysis PCA with heterogeneous data.)

The concept of random variable can be extended to that of a *random vector* $\mathbf{X}$ in dimension pD in the standard way, by adding more components in a Cartesian way. Likewise for the mean, variance and standard deviation of a random vector $\mathbf{X}$.

For a 2D discrete random vector $\mathbf{X} = (X, Y)$, the *joint pdf* of X is given by $f_{X,Y}(x, y) = P(X = x, Y = y)$ and its marginal pdf's are

$$f_X(x) = P(X = x) = \sum_y P(X = x, Y = y) = \sum_y f_{X,Y}(x, y), \text{ and}$$

$$f_Y(y) = P(Y = y) = \sum_x P(X = x, Y = y) = \sum_x f_{X,Y}(x, y).$$

11.1.1 Commonly Used Discrete Distributions

Table 11.2 shows common discrete probability distributions and typical applications.

Discrete Uniform Distribution
$X \sim U(1, 2, ..., m)$ with pdf $f(x) = \frac{1}{m}$, $x = 1, 2, \cdots, m$ $(m \geq 1)$.

One random experiment for this distribution is to toss a fair die once and observe for X the number shown. In this case, $X \sim U(1, 2, ..., 6)$. Another experiment for such a distribution is to choose a random digit and observe for X the digit chosen. In this case, $X \sim U(0, 1, 2, ..., 9)$.

1. $E(X) = \frac{m+1}{2}$.
2. $Var(X) = \frac{m^2-1}{12}$.

In the general case, choosing an integer at random in the interval $[A, B]$ for X gives $X \sim U(A, A + 1, ..., B)$ with pdf $f(x) = \frac{1}{C}$, $x = A, A + 1, \cdots B$, where $C = (B - A + 1)$.

1. $E(X) = \frac{A+B}{2}$.
2. $Var(X) = \frac{C^2-1}{12}$, where $C = (B - A + 1)$.

Binomial Distribution

$X \sim B(n, p)$ with pdf $f(x) = \binom{n}{x} P^x (1 - P)^{n-x} = \frac{n!}{x!(n-x)!} P^x (1 - P)^{n-x}, \quad x = 0, 1, \cdots, n.$

One experiment for such a distribution is to toss a fair coin 10 times and observe for X the number of successes (say, heads.) In this case, $X \sim B(10, 0.5)$.

1. $E(X) = nP.$
2. $Var(X) = nP(1 - P).$

There are conditions on the random experiment to yield a RV with this distribution:

- the experiment consists of n identical and independent trials;
- each trial is Bernoulli, i.e., it results in one of two outcomes, success or failure;
- from trial to trial, the probability of success on a trial is P and remains constant. The probability of failure is $1 - P$.

Poisson Distribution

$X \sim Poisson(\lambda)$ with pdf $f(x) = \frac{\lambda^x e^{-\lambda}}{x!}, \quad x = 0, 1, 2, \cdots$

One experiment for such a distribution is to observe for X the number of traffic accident at a certain location during a certain time interval (e.g., rush hour.)

1. $E(X) = \lambda.$
2. $Var(X) = \lambda.$

Geometric Distribution

$X \sim Geo(P)$ with pdf $f(x) = P(1 - P)^x, \quad x = 0, 1, 2, \cdots$

One experiment for such distribution is to observe for X the number of (repeated) failures in a Bernoulli trial needed to obtain the first success.

1. $E(X) = \frac{1-P}{P}.$
2. $Var(X) = \frac{1-P}{P^2}.$

Negative Binomial Distribution

$X \sim NB(r, P)$ with pdf

$$f(x) = \binom{x + r - 1}{r - 1} P^r (1 - P)^x, \quad x = 0, 1, 2, \cdots$$

One experiment for a such distribution is to observe for X the number of failures in a Bernoulli trial until the rth success is observed.

1. $E(X) = r\frac{1-P}{P}.$
2. $Var(X) = r\frac{1-P}{P^2}.$

Hypergeometric Distribution

A RV X has a hypergeometric distribution if and only if its pdf is

$$f(x) = \frac{\binom{S}{x}\binom{N-S}{n-x}}{\binom{N}{n}},$$

1. $E(X) = n\frac{S}{N}$.
2. $Var(X) = n\frac{S}{N}\frac{N-S}{N}\frac{N}{N-1}$.

One experiment for such distribution is to draw balls from an urn and for X to observe the number of balls drawn marked as "Success" in choosing n balls without replacement from an urn containing N balls, with s of them marked as "Success" and the remaining $N - s$ balls marked as "Failure."

11.1.2 Commonly Used Continuous Distributions

Table 11.3 shows commonly arising continuous probability distributions and their characterizing parameters.

Uniform Distribution

$X \sim U(A, B)$ with pdf $f(x) = \frac{1}{B-A}$, $A \leq x \leq B$.

1. $E(X) = \frac{A+B}{2}$.
2. $Var(X) = \frac{(B-A)^2}{12}$.

The special case with $A = 0$, $B = 1$ yields the standard uniform continuous distribution $X \sim U(0, 1)$ in the interval $[0, a]$ with pdf

$$f(x) = 1, \quad 0 \leq x \leq 1.$$

1. $E(X) = \frac{1}{2}$.
2. $Var(X) = \frac{1}{12}$.

Normal Distribution

$X \sim N(\mu, \sigma^2)$ with pdf $f(x) = \frac{1}{\sqrt{2\pi}\sigma}e^{-(x-\mu)^2/(2\sigma^2)}$.

1. $E(X) = \mu$.
2. $Var(X) = \sigma^2$.

If $X \sim N(\mu, \sigma^2)$, then $Y = aX + b \sim N(a\mu + b, a^2\sigma^2)$.

The standard normal distribution is $Z \sim N(0, 1)$ with pdf $f(z) = \frac{1}{\sqrt{2\pi}}e^{-z^2/2}$.

1. $E(Z) = 0$
2. $Var(Z) = 1$

Logistic Distribution

$X \sim Logistic(\mu, \beta)$ with pdf $f(x) = \frac{1}{\beta}\frac{e^{-(x-\mu)/\beta}}{(1+e^{-(x-\mu)/\beta})^2}$.

1. $E(X) = 0$.
2. $Var(X) = \frac{\pi^2}{3\beta^2}$.

The standard logistic distribution is $Logistic(0, 1)$ with $\mu = 0$ and $\beta = 1$, i.e., its pdf is $f(x) = \frac{e^x}{(1+e^x)^2}$, or equivalently, $f(x) = \frac{e^{-x}}{(1+e^{-x})^2}$, $-\infty < x < \infty$..

This is an important distribution because the popular logistic regression is based on this distribution.

Exponential Distribution
$X \sim Exp(\lambda)$ with pdf $f(x) = \lambda e^{-\lambda x}$, $x \geq 0$.

1. $E(X) = \frac{1}{\lambda}$.
2. $Var(X) = \frac{1}{\lambda^2}$.

The standard exponential distribution is $Z = \lambda X$, with pdf $f(z) = e^{-z}$, $z \geq 0$.

1. $E(Z) = 1$.
2. $Var(X) = 1$.

Double Exponential Distribution
$X \sim DExp(\lambda)$ with pdf $f(x) = \frac{\lambda}{2} e^{-\lambda|x|}$.

1. $E(X) = 0$.
2. $Var(X) = \frac{2}{\lambda^2}$.

Gamma Distribution
$X \sim Gamma(\alpha, \beta)$ with pdf $f(x) = \frac{1}{\Gamma(\alpha)\beta^\alpha} x^{\alpha-1} e^{-x/\beta}$, $0 \leq x < \infty$.

1. $E(X) = \alpha\beta$.
2. $Var(X) = \alpha\beta^2$.

Beta Distribution
$X \sim Beta(\alpha, \beta)$ with pdf $f(x) = \frac{1}{B(\alpha,\beta)} x^{\alpha-1}(1 - x)^{\beta-1}$,

where $0 < x < 1$ and $B(\alpha, \beta) = \frac{\Gamma(\alpha)\Gamma(\beta)}{\Gamma(\alpha+\beta)}$ is a complete beta function.

1. $E(X) = \alpha\beta$.
2. $Var(X) = \alpha\beta^2$.

χ^2 Distribution
$X \sim \chi^2(v)$ if and only if $X \sim Gamma(v/2, 2)$.

1. $E(X) = v$.
2. $Var(X) = 2v$.

When $v = 2$, $X \sim \chi^2(2)$ if and only if $X \sim Exp(2)$.
If $U \sim U(0, 1)$, then $X = -2\ln(U) \sim \chi^2(2)$.
If $Z \sim N(0, 1)$, then $X = Z^2 \sim \chi^2(1)$.

Student's t Distribution

$X \sim t(v)$ with pdf $f(x) = \frac{\Gamma\left(\frac{v+1}{2}\right)}{\Gamma\left(\frac{v}{2}\right)\sqrt{v\pi}}\left(1 + \frac{x^2}{v}\right)^{-\frac{v+1}{2}}, \quad -\infty < x < \infty.$

1. $E(X) = 0$ (where $v > 1$.)
2. $Var(X) = \frac{v}{v-2}$ (where $v > 2$.)

F Distribution

Y with pdf $Y = \frac{X_1/v_1}{X_2/v_2}$, where $X_1 \sim \chi^2(v_1)$ is independent of $X_2 \sim \chi^2(v_2)$.

1. $E(X) = \frac{v_2}{v_2-2}$ (where $v_2 > 2$.)
2. $Var(X) = 2\left(\frac{v_2}{v_2-2}\right)^2 \frac{v_1+v_2-2}{v_1(v_2-4)}$ (where $v_2 > 4$.)

The F distribution is commonly used in ANOVA tests.
Connection with the t distribution: if $X \sim t(v)$, then $Y = X^2 \sim F(1, v)$.
Connection with the beta distribution: if $Y \sim F(v_1, v_2)$, then $Y = \frac{v_2}{v_1}\frac{X}{1-X}$, with
$\quad X \sim Beta(v_1/2, v_2/2)$.

Weibull Distribution

X with pdf $f(x) = \frac{\alpha}{\beta}x^{\alpha-1}e^{-x^\alpha/\beta}, \quad 0 \le x < \infty.$

Cauchy Distribution

X with pdf $f(x) = \frac{1}{\pi b\left(1 + \left(\frac{x-a}{b}\right)^2.\right)}$

1. $E(X)$ does not exist.
2. $Var(X)$ does not exist.

Multivariate Normal Distribution

$X \sim N_p(\mu, \Sigma)$ with pdf $f(\mathbf{x}) = \frac{1}{\sqrt{|2\pi\Sigma|}}\exp\left(-\frac{(\mathbf{x}-\mu)'\Sigma^{-1}(\mathbf{x}-\mu)}{2}\right)$
$\quad$ If $\mathbf{X} \sim N_p(\mu, \Sigma)$, then $\mathbf{Y} = \mathbf{AX} + \mathbf{b} \sim N_p(\mathbf{A}\mu + \mathbf{b}, \mathbf{A}\Sigma\mathbf{A}')$
$\quad$ If $\mathbf{Z} = \Sigma^{-1/2}(\mathbf{X} - \mu) \sim N_p(\mathbf{0}, \mathbf{I})$, then $\mathbf{Z} \sim N_p(\mathbf{0}, \mathbf{I})$ and

$$\mathbf{X} = \mu + \Sigma^{1/2}\mathbf{Z} \sim N_p(\mu, \Sigma).$$

The marginal distribution and conditional distribution are normal distributions.

The well-known property that $\left(\frac{X-\mu}{\sigma}\right)^2 \sim \chi_1^2$, where $X \sim N(\mu, \sigma^2)$, can be extended to

$$(\mathbf{X} - \mu)'\Sigma^{-1}(\mathbf{X} - \mu) \sim \chi_p^2$$

Trinomial Distribution (Extension of the Binomial Distribution)

If there are three possible outcomes for a trial, where X is the count for the first category with probability p_x, Y is the count for the second category with probability p_y and n is the total number of trials, $Z = n - X - Y$ is the count for the third category with probability $1 - p_x - p_y$. This random vector $\mathbf{X} = (X, Y)$ follows a trinomial distribution with the probability density function

$$P(X = x, Y = y) = \frac{n!}{x!y!(n - x - y)!} P_x^x P_y^y (1 - P_x - P_y)^{n-x-y},$$

where $x \geq 0$, $y \geq 0$ and $x + y \leq n$ This is an extension of the binomial distribution where the marginal distributions of X, Y, and $X + Y$ are $B(n, p_x)$, $B(n, p_y)$, $B(n, p_x + p_y)$, respectively.

It can be further extended to multinomial distribution with more categories.

11.1.3 Major Results in Probability and Statistics

Bayes' Theorem
If $A, B \subset \Omega$ are events in a probability space with $P(A)P(B) \neq 0$, then

$$P(B \mid A) = P(A \mid B)P(B)/P(A).$$

More generally, if $\Omega = B_1 \cup B_2 \cup \cdots \cup B_m$ is a partition of the sample space Ω (hence $B_i \cap B_j = \emptyset$ for $i \neq j$), with $P(B_i) \neq 0$, then

$$P(B_i \mid A) = P(A \mid B_i) \, P(B_i)/[\Sigma_i P(B_i)P(A \mid B_i)].$$

For the next three results, let $X_i, i = 1, 2, \cdots, n$ be a random sample of size n from a general distribution with mean μ and variance σ^2 and let $\bar{X}_n = \frac{1}{n} \sum_{i=1}^n X_i$ be the mean of the initial subsample of size n.

Central Limit Theorem
If the sample size n is large, regardless of the given distribution of the X_is, the random variable $\bar{X}$ is approximately normally distributed

$$\bar{X} \sim N(\mu, \sigma^2/n), \quad \frac{\bar{X} - \mu}{\sigma/\sqrt{n}} \sim N(0, 1).$$

That is, the limiting distribution of $W_n = \frac{\bar{X} - \mu}{\sigma/\sqrt{n}}$ will converge to $N(0, 1)$ as $n \to \infty$.

Chebyshev's Inequality
Regardless of the distribution of the X_is, for every $\epsilon > 0$,

$$P(\mid \bar{X} - \mu \mid \geq \epsilon) \leq \frac{\sigma^2}{n\epsilon^2}.$$

Law of Large Numbers

Regardless of the distribution P and the value of $\epsilon > 0$, the distribution of $\bar{X}$ is concentrated around its mean μ, i.e.,

$$P(|\bar{X}_n - \mu| > \epsilon) \to 0,$$

for all $n > N_\epsilon$, for some N_ϵ sufficiently large.

11.2 Linear Algebra Background

This section summarizes important facts in linear algebra and matrices that are useful to understand their role and implications for data science.

11.2.1 Fields, Vector Spaces and Subspaces

Concepts such as numbers of various kinds (natural, integer, rational, real and perhaps complex) are usually part of the modern high school curriculum. The concept of a *vector* occurs in courses like physics and chemistry as a quantity that has associated with it not only a real number as a length but also a direction, such as force, velocity and acceleration, usually represented by arrows to signify magnitude (the length) and direction (the tip of the arrow).

These concepts are generalized in Cartesian geometry by abstraction into *scalars* and *vectors* in higher dimensions beyond 3D using the same intuitions (just like distances are in Sect. 1.1) and even into more abstract objects, still called *scalars* and *vectors* by analogy. Mathematicians have produced a systematic exploration of these concepts in the subfield of linear algebra, by distilling out the main concepts necessary for working with vectors and coordinate systems in a more general and effective fashion. The key rules to require are the usual properties of real and complex numbers for the scalars and vectors for vectors spaces. The advantage of this abstract setup is that the thinking only has to be done once by mathematicians, but conclusions can be applied to any set of objects by us all, just as long as they satisfy some key rules, as described next. They can be reasoned with in just the same way one would with lengths and arrows pointing in some direction, even though they may not be them literally.

A *field F* is a set of objects endowed with two operations addition (+) and multiplication (·) between pairs of them that satisfy the usual properties of their likes with real numbers in **R**:, i.e., for all elements $a, b, c \in F$:

- Both operations are *associative* $((a+b)+c = a+(b+c))$, *commutative* $(a+b = b+a$ and $a \cdot c = c \cdot a)$, have *neutral elements* (like 0 for addition and 1 for multiplication) and *inverses* (denoted $-a$ for addition and b^{-1} for multiplication, except for 0, so that $a + 0 = a$, $a + (-a) = 0$ and $b \cdot b^{-1} = 1$ for $b \neq 0$.)

- Multiplication distributes over addition, i.e.,

$$a \cdot (b + c) = (a \cdot b) + (a \cdot c).$$

(The $\cdot$ for multiplication is, as usual, dropped to ease notation.)

Example 11.2 The set $\mathbf{R}$ of real numbers is a field with the usual operations of addition and multiplication. So are the sets of complex and of rational numbers. Even the set with just two elements $\mathbf{B} = \{0, 1\}$ (the Boolean set) is a field under addition modulo 2 and multiplication as usual, which makes it the smallest field possible. It is even interesting to note that the hours in a clock of only 5 (or 3 or 7, or any prime number of) hours, i.e., the set $\mathbf{Z}_5 = \{0, 1, 2, 3, 4\}$ is also a field if the operations are performed as usual with integers, but taking care of subtracting or adding 5 every time a result goes out to bring it back to this range 0-4 (e.g., $2+3 = 0$ and $2 \cdot 3 = 6 - 5 = 1$ so 2,3 are both additive and multiplicative inverse of one another. These latter are examples of finite fields (know as *Galois fields*, in honor of their discoverer, the French mathematician E. Galois [9] in the 1800s.) ☐

The concept of vector space requires a scalar field F. In this book, as in common applications, the field of scalars is always the real $\mathbf{R}$ or complex numbers $\mathbf{C}$.

A *(linear) vector space V over a field F* (with elements called *scalars*) is a nonempty set of objects called *vectors* endowed with two operations of addition (+) and multiplication by scalars from F (indicated by just concatenation) that satisfy the usual properties of their likes with real numbers in $\mathbf{R}$ and vectors in Cartesian spaces, i.e., for all elements $\lambda, \mu \in F$ and $u, v, w \in V$:

- addition is *associative* $((u + v) + w = u + (v + wc))$, *commutative* $(u + v = v + u)$, has a *neutral element* (like 0 for addition) and *inverses* (denoted $-u$) so that $u + (-u) = u - u = \mathbf{0}$;
- Multiplication by a scalar is also *associative* and *distributes over* vector addition, i.e.,

$$\lambda(\mu u) = (\lambda \mu)u \text{ and } \lambda(u + v) = \lambda u + \lambda v.$$

- Scalar multiplication by the multiplicative identity scalar leaves vectors unchanged, i.e., $1v = v$ for all $v \in V$.

Example 11.3 If $d > 0$ is an integer and F is any field, the Cartesian product $V = F^d = F \times \cdots \times F$ (d times) is a vector space over F with vector addition and multiplication defined componentwise. (To ease the notation, a column vector is written as the transpose (x') of a row vector $x = (x_1, \cdots, x_d)$.) The familiar vector space $\mathbf{R}^d$ (like the 2D plane and 3D spaces) are thus recovered as particular cases with the real number field. ☐

A subset W of a vector space V is a *subspace* if it satisfies the properties required by the definition of vector spaces with the operations inherited from the container space V, i.e., W is closed under addition, taking additive inverses and multiplying its elements by arbitrary scalars.

A subspace is never empty, as it must contain at least the neutral element $\mathbf{0}$ under addition.

Example 11.4 If $F = \mathbf{B}$, the smallest field (in fact, the Galois field $\mathbf{Z}_2$), the vector space of dimension d over $\mathbf{B}$ is called the (Boolean) *hypercube* of dimension d and consists of binary vectors of length/dimension d. It plays an important role in Shannon's theory of communication and error-detecting and error-correcting codes for robust communication through a noisy channel.   □

11.2.2 Linear Independence, Bases and Dimension

The following concepts make sense in an arbitrary vector space V over a field F (which can be thought of as $\mathbf{R}$ for the purposes of this book.)

A *linear combination* in V is a vector obtained as a sum of scalar products

$$\lambda_1 v_1 + \lambda_2 v_2 + \ldots + \lambda_n v_n$$

of some finite subset of elements $v_1, v_2, \ldots, v_d \in V$ with scalars $\lambda_1, \lambda_2, \ldots, \lambda_n \in F$, The set of all possible linear combinations of vectors $v_1, v_2, \ldots, v_n \in S$ is the (linear) *span* of a subset $S \subseteq V$ and it is denoted $\mathscr{L}(S)$.

The subset S is *linearly independent* if no element in it is a linear combination of any other elements in S. It is called *linearly dependent* otherwise.

Example 11.5 The vectors $u = (1, 2, 3)'$, $v = (2, -1, 4)'$ and $w = (0, -5, -2)'$ are linearly dependent since $2u - v + w = 0$, i.e., $v = 2u + w$. A set containing the zero vector $\mathbf{0}$ is always linearly dependent as $\mathbf{0}$ can always be obtained from others as a linear combination with scalar coefficients 0.   □

Example 11.6 The vectors $u = (1, 1, 1)'$ and $v = (1, 0, -1)'$ are linearly independent because neither one is a linear combination (in this case, that means a scalar multiple) of the other: $\lambda u = (\lambda, \lambda, \lambda) = v = (1, 0, -1)$ implies that $\lambda = 1 = -1 = -\lambda$, which is nonsense (except perhaps in the Galois field $\{0, 1\}$.) If a linear combination is 0, e.g., $\lambda u + \mu v = 0$ for some $\lambda, \mu \in \mathbf{R}$, then $(\lambda + \mu, \lambda, \lambda - \mu)' = 0$ and this would mean that both λ and μ are the scalar 0, the trivial linear combination producing the vector $\mathbf{0}$.   □

It is evident that $\mathscr{L}(S)$ is always a (linear) subspace of V. For a vector space V, it is of much interest to find a smallest subset B that generates the *full* space by doing linear combinations with scalars from F.

A (linear) *basis* of a vector space V is a subset of vectors B with the smallest number of elements that spans the space V, i.e., $\mathscr{L}(B) = V$. The (linear) *dimension* of V is the cardinality (number of elements) of such a basis.

Such a set has to be linearly independent since a vector that is a linear combination of others can always be excluded to obtain a smaller such set with the same linear span. Naturally, a vector space V that has a finite basis B is called *finite dimensional*. Not all vector spaces are finite dimensional (for example, the space consisting of one-way infinite sequences $(x_1, x_2, \cdots)$ of real numbers with componentwise operations. Such spaces will be of no concern in this book since they are out of reach for conventional computers, unless finite dimensional approximations are used.)

Example 11.7 The vector space $\mathbf{R}^d$ is finite dimensional since it has a basis $B = \{e_1, e_2, \ldots, e_d\}$ where $e_i = (0, \ldots, 1, \ldots, 0)'$ has 0 coordinate everywhere except at the ith position, where $e_i = 1$. It is easy to verify that every vector $v = (v_1, v_2, \ldots, v_d)'$ is a linear combination of the basis elements $v = v_1 e_1 + v_2 e_2 + \ldots + v_d e_d$. This basis is called the *standard basis*, so the dimension of $\mathbf{R}^d$ is $dim(\mathbf{R}^d) = d$. □

A subspace always has dimension no larger than that of the entire host space V, although this is not entirely obvious from the definitions above.

A subspace W of a finite dimensional vector space of finite dimension is never larger than $dim(V)$, i.e. $dim(W) \leq dim(V)$.

Example 11.8 Given a basis B of a vector space V, every vector must be expressible as a linear combination of the elements of B. The coefficients in this combination are *unique*, because two different combinations being equal would lead to a combination for **0** (by transposing terms to just one side and re-grouping) with some nonzero coefficient(s), which would in turn lead to one vector being expressible as a linear combination of the others, nonsense since elements in B are linearly independent. These coefficients can be put together into a vector $(\lambda_1, \lambda_2, \cdots, \lambda_d)$ to obtain a Cartesian object over the scalar field so one can get rid of the abstract nature of the objects in the original vector spaces V and handle them as usual business in $\mathbf{R}^d$ for a vector space of dimension d, or in F^d for some scalar field F. □

11.2.3 Linear Transformations and Matrices

In a linear space, as the basis elements go, so follows the rest of the space, as illustrated by the following facts.

A *linear transformation* (or *mapping*) $T : V \rightarrow W$ between two vector spaces over the same field F is a function that preserves linear combinations, i.e, for every pair of scalars $\lambda, \mu \in F$ and of vectors $u, v \in V$,

$$T(\lambda u + \mu v) = \lambda T(u) + \mu T(v).$$

Example 11.9 In the 2D Euclidean space, any rotation counterclockwise (say 45°) is a linear transformation. A perpendicular projection onto a 1D line (such as the 45° line) is also a linear map. However, a translation by a fixed nonzero vector v_0 like $(1, 1)$ given by $T(u) = u + v_0$ is not since $T(u + v) = u + v + v_0 \neq (u + v_0) + (v + v_0) = (u + v) + 2v_0$. (The latter are called *affine* transformations.) □

There are several interesting subspaces associated with a linear map.

The *kernel* (or *null space*) K of a linear map $T : V \to W$ consists of all vectors in V that collapse to the $\mathbf{0}$ of W under T, i.e.,

$$K = \{u \in V | T(u) = \mathbf{0}\} = T^{-1}(\mathbf{0}).$$

The *range* of T is the set $T(V) = \{T(u) : u \in V\}$, i.e., the set of all images $T(u)$ obtained by ranging u through all elements of V.

The action of any linear map is entirely determined by its mapping of the elements in any basis of V. Linear maps can be represented by yet a richer concept of a number over a given field F.

A *matrix* A (or $A_{d \times k}$ if dimensions are important) over a field F is a rectangular array of dk elements of F (d rows and k columns.) The matrix A *associated to a linear map* T for given bases B, B' of V, W respectively, has as i^{th} column (out of dim(V) $= d$ columns) of dimension $k = dim(W)$ consisting of the coefficients of the image of the ith element b_i in B with respect the basis B'; i.e., $A = [T(b_1)', \ldots, T(b_d)']$.

Such a matrix representation for the linear transformation T is *not unique* as it depends on the choice of bases B, B', although it is of course unique if the bases are fixed. Once B, B' are fixed, it can serve as a full representation of T. For example, one advantage is that the image of an arbitrary element $u \in V$ can be obtained as

$$T(u) = Au',$$

i.e., just as the *matrix product* of A and the column vector u' with coefficients the components of u with respect to the basis B. Thus, just like a linear function in one variable is given by $f(x) = mx$, where m is the slope of the line representing its graph, so is the concept of matrix A an abstraction of the concept of slope to a generalized vector space. In particular, one can regard a given dataset (a 2D table) as a matrix A defining some linear transformation and whose decomposition(s) below provide some high-level analysis of the dataset (examples are found in Sects. 2.1 and 9.3.)

11.2.4 Eigenvalues and Spectral Decomposition

A lot of other essential information about T is hidden inside A.

An *eigenvalue* of A (and so of T) is a scalar λ such that $T(u) = \lambda u$ for some nonnull vector u, i.e., T only stretches (if $|\lambda| > 1$) or shrinks (if $|\lambda| < 1$) u by a factor of λ (perhaps reversing orientation if $\lambda < 0$) without changing its direction Such a vector u is called an *eigenvector* of T for the eigenvalue λ.

Therefore, if T happens to have a set of linearly independent eigenvectors that form a basis of V, understanding the effect of the transformation T becomes evident.

Example 11.10 If the eigenvectors are e_1 and e_2 in the 2D Euclidean plane and T has two eigenvalues $\lambda_1 = 2$ and $\lambda_2 = -1$, the vector $v = (3, 4)$ decomposes as

$V = 3e_1 + 4e_2$, and then the effect of T on v is $T(v) = 3T(e_1) + 4T(e_2) = 3(2e_1) + (-1)e_2) = 6e_1 - 4e_2$, i.e., to stretch the e_1 component by a factor of 2 and the e_2 component by a factor of (-1) (reversing orientation) without stretching since $|\lambda| = |-1| = 1$ and combine the results back to get $T(v) = (6, -4)$. If the eigenvectors happen to be rotated with respect to the standard basis, changing the basis to a basis of eigenvectors will simplify the expression of T into a just a diagonal matrix (where the entries are zero (0) off the main diagonal (entries $A_{ij} = 0$ where $i \neq j$.) □

The eigenvectors of a linear transformation $T: V \to V$ within a space V can be computer manually by solving a system of equations determined by $T(u) = \lambda u$, for each eigenvalue λ root of the polynomial equation of degree d vanishing the determinant

$$|I - \lambda A| = 0,$$

one equation for each component, where I is the identity matrix of the same dimension. Alternatively, they can be computed (perhaps only approximately) by using software libraries in a computational platform (some are described in Sect. 11.6.)

Spectral Decomposition
If $A = A_{d \times d}$ has d different eigenvalues $\lambda_1, \ldots, \lambda_d$ and their eigenvectors are linearly independent, then A has a factorization (the spectral decomposition of A) of the form

$$A = Q \Lambda Q^{-1},$$

where Q is an invertible matrix consisting of the eigenvectors of A (as columns) and Λ is a diagonal matrix with the eigenvalues in the main diagonal (i.e., all entries off the diagonal are 0, hence denoted just by $\Lambda = \mathbf{D}(\lambda_1, \ldots, \lambda_d)$, or just $(\lambda_1, \ldots, \lambda_d)$.)

Example 11.11 For the transformation in Example 11.10, $\Lambda = \mathbf{D}(2, -1)$ and the columns of Q are $u_1 = (1, 0)'$ and $u_1 = (0, 1)'$, i.e., $Q = I$ is the identity matrix (which it its own inverse) and A is its own spectral decomposition $A = \Lambda$. If T is a rotation as in Example 11.9, then the matrix of T with respect to the standard basis is $A - [s(1, 1)' \quad s(-1, 1)']$, where $s = \sin(45°) = \cos(45°) = \sqrt{2}/2$ is a shrink factor to get the rotation of the standard vector $(1, 1)'$ to have length 1, i.e., $s = \sqrt{2}/2$. The eigenvalues of A are the solution to the quadratic equation

$$| I - \lambda A | = [(1, 0)'(0, 1)' - \lambda A | = \lambda^2 - s\lambda + (1 + s^2),$$

which are not real, but have a complex imaginary component, so A is not diagonalizable. □

All symmetric matrices have real eigenvalues (possibly with so-called *multiplicities* when two eigenvalues are equal, as in the solution of $(\lambda - 1)^2 = 0$, in which case the eigenspace generated by the eigenvectors of such an eigenvalue have dimension larger than 1.) Nonetheless, in the more general case of a nondiagonalizable matrix, a similar decomposition is possible if one is willing to substitute the entries in the spectral decomposition by *square blocks* submatrices (of dimension equal to the dimension of the eigensubspaces corresponding to such eigenvalues of multiplicity larger than 1.)

In Euclidean spaces, every matrix A has a Jordan Canonical Form (*or* Jordan *decomposition*) *as a product*

$$A = P \Lambda Q ,$$

where Λ is a block superdiagonal matrix with eigenvalues in the tridiagonal band and possibly some ones (1s) in the first superdiagonal (entries λ_{ij} with $j = i + 1$.)

In the most general case where the range space W of T has different dimension from V, a decomposition is still possible, but the matrices may no longer be square, e.g., Λ may be a truncated diagonal matrix and P, Q must have the appropriate dimensions for the matrix products to make sense (in a product ΛQ, the number of columns of Λ must match the number of rows of Q.) This factorization is called the *Singular Value Decomposition (SVD)* of A. This decomposition has found many interesting applications in data science, for example in Latent Semantic Analysis (LSA), where the number of documents (data points) $n = d$ is different from the number of distinct words (features) p appearing in all the documents. (More details can be found in Sects. 4.3 and 10.2.)

Finally, Euclidean and Boolean spaces are yet richer in structure. The length of the projection of a vector u onto another v can be computed using the so-called *dot product* (also *scalar product*), given by

$$\gamma = u \cdot v = \sum_{i=1}^{d} u_i \, v_i = \|u\| \|v\| \cos(\alpha)$$

(sometimes simply denoted by the juxtaposition uv.) The projection of u onto v is then γv (and also γu of v onto u.) So, vectors orthogonal to one another have dot product 0 and a matrix A is *orthogonal* exactly when every pair of its column vectors are orthogonal. They represent linear transformations obtained by rotating the standards basis vectors e_i to other vectors (the columns in A) so as to preserve their orthogonality between any pair of them. In particular, the *length* of a vector (distance of the tip from the origin $\mathbf{0}$) is then given by $\|u\| = \sqrt{u \cdot u}$. This structure applies as well to Boolean hypercubes, which are metric spaces with the resulting Hamming distance between u, v given by $Ham(u, v) = \sum_{i=1}^{d} u_i \oplus v_i$, the sum of the XOR (exclusive *or*) of their components.

Further details can be found in any standard textbook in linear algebra, e.g., [20].

11.3   Computer Science Background

This section summarizes important facts in computer science that are useful to understand its role and implications for data science, as we well as its scope and limitations.

11.3.1   Computational Science and Complexity

The working definition of data science is problem solving to extract information about questions concerning a population from samples of it. To really make sense of this definition, the question of *What exactly is information?* would need to be addressed. That is precisely the subject matter of computer science and so it plays a central role in any science, data science in particular.

The subject matter of computer science would appear to go along the lines of *the science concerned with the study of computers*. This definition turns out to be inadequate for a variety of reasons. Among many others, there is a difference to be made between information and its embodiment, usually called *data*, as well as between information processing and *data communication* (which is what Shannon's concept of information aims to answer.) Second, there is no other science like it (like car science, or microscope science, or even human science; but there is computer science!) Third, a case can also be made that there are different kinds of information, such as *knowledge* and even *wisdom*. The funny thing is, humans and living organisms alike are all remarkable users of information, yet no one can define precisely what it is. Thus, defining information is a far more difficult challenge that has remained and will probably remain unsolved for a good time to come [5].

Another way to define computer science would be along the lines of *the study of information and its processing*. But alas, in addition to answering "What is *information?*", this definition would require an answer to the question "What is a *processor*." Therefore, it is not surprising that the digital age has been born more directly concerned with data processing *devices*, particularly in their specific electronic implementation, i.e., the conventional computers of our time.

Yet another way to characterize a science is by defining the scope or aspect of the world it is concerned with. Like data science, the scope of computer science can then be regarded as concerning computational problems in the sense that they all call for certain types of inputs as data (as described in Sect. 1.3) to a problem and spell out an expectation in terms of the kind of result to be produced as an answer.

> An *algorithmic problem* (AP) is a list of questions, each fully expressible as a finite string over a fixed finite alphabet of characters, with an answer determined by the question and also expressible as a finite string over the same alphabet.

APs include all data science problems dealt with in this book, but can be much more general in terms of the questions to be answered and the solutions provided.

Example 11.12 Primality testing is one of the most representative examples. The alphabet is decimal (another problem could be in binary) and questions are follows.

[REC(Primes)]   **PRIMALITY TESTING**

> INSTANCE: A decimal string x
> QUESTION: Does x represent a prime integer (no proper factorization)?

There is an infinite number of questions (one for each x, e.g., $x = 19$ or 32), but for each there is a unique answer (e.g., *Yes*, *No* respectively.)

On the other hand, the outcome of a given toss of a specific coin (heads or tails) is not an AP because first, it is probably impossible to fully specify a coin with a finite string containing full information to determine it as a physical object different from other coins, but more importantly, the answer to a question is not unique and fully determined by the coin (even for the same coin) in advance.   □

APs come in three flavors, *recognition problems*, *generation problems* (how to produce all and only the strings in a given language, like Python 7.9, from some specification of it) and *compaction problems* (how to produce finite strings fully describing potentially infinite sets such as a certain natural or computer programming language.) Recognition problems have proved challenging enough that most of what is understood by the foundations of computer science today is concerned with them. Primality testing **[REC(Primes)]** is a typical example. A recognition problem is specified by a set of strings (called a formal *language*) L over some finite alphabet Σ as follows.

[REC(L)]   **MEMBERSHIP**(L)

> INSTANCE: A string x over alphabet Σ
> QUESTION: Does x belong to L ("Yes" or "No")?

Of course, many other types of APs are of interest in their own right in practice. Common problems are **[SORTING]** and **[SEARCHING]**. A particular type is particularly relevant to data science because they are about optimization of the type involved in machine learning and data science algorithms.

[TSP]   **[TRAVELING SALES PERSON]**

> INSTANCE: A weighted finite graph G consisting of vertices V and (undirected) edges joining them, weighted by nonnegative integer numbers
> QUESTION: What is a tour of the graph visiting all nodes with the shortest total sum of (the) weight(s) among all possible tours?

[SAT]   **[SATISFIABILITY]**

> INSTANCE: A Boolean formula $\varphi(x_1, \cdots, x_k)$ of k Boolean variables
> QUESTION: Is φ satisfiable, i.e., does it evaluate true for any one assignment of truth values of its variables $x_1, \cdots, x_k$?

For example, the formula $\varphi(x, y) = (x \, and \, y)$ is satisfiable (if both x, y are assigned the value *True*), but $\varphi(x, y) = (x \, and \, \bar{x})$ is not ($\bar{x}$ is the Boolean negation of x.)

[**MORT**(d)]   [**MORTALITY**](d)

INSTANCE: A finite set S of matrices of size $d \times d$ with integer entries
QUESTION: Is the set S mortal?

Here, a set of matrices is *mortal* if some product of elements from S (in some order and possibly with repetitions of the choices) multiply out to the **0** (null) matrix. Although with numbers ($d = 1$) this is only possible if some element in S is 0 itself, as generalized numbers, products of two nonzero square matrices of size 2×2 may actually be the null matrix (like, two diagonal matrices with a single 1 in their diagonal at different positions.)

What constitutes an acceptable solution of an arbitrary AP *in general*? Intuitively, one would expect some kind of device that answers the questions in the AP, i.e.,

[S1] the device must be able to read any of the questions x in the problem as an input, work on it for a *finite* period of time and eventually stop (halt) the processing, after
[S2] returning the *true answer* to question x; in addition,
[S3] the same device must do so systematically for every question x in the AP without any further modification, after its design is complete.

A *solution* to an AP is usually some sort of model or program/code, ultimately to be translated into some sort of physical device to be run, typically a conventional computer, but possibly very different (e.g., a human brain or a quantum computer) that satisfies conditions [S1]–[S3] above.

What constitutes an acceptable specification of a device (which may quickly turn into wishful thinking for a magical solution) is beyond the scope of this book. (Alan Turing's proposal in the foundational paper that marks the birth of computer science [21] in 1936 is considered to be the simplest and rigorous standard.) Nevertheless, one can get an understanding of what is now called a *Turing machine* with an intuitive equivalent, the concept of an *algorithm*. Computer scientists generally agree that the following thesis is the appropriate characterization of what constitutes an acceptable solution to an AP, arrived at after searches lasting over half a century of computer science by computer scientists all over the world for a sound, objective and appropriate definition of "computer" and discovering that many other possibilities are just Turing machines in disguise. That includes concepts such as algorithms, computer programs, mechanical procedures, automatic procedures, automatable procedures and many others.

The Church–Turing Thesis
Every algorithm solving an algorithmic problem can be specified and implemented as a Turing machine.

In a nutshell, an algorithmic problem is solvable if some device exists (which is not quite to say it will be easy for humans to find) implementing some algorithm that receives input data encoded as strings, crunches symbols with a *fixed program but unbounded storage and running time*, and returns the answer encoded in a finite string expressing the answer to *each* and *all* question(s) in the AP, correctly and systematically. Solving an AP requires either a full and rigorous description of a device that must ideally satisfy the three conditions [S1]–[S3], or, alternatively, some argument that such a device could not possibly exist (i.e., that the problem is *unsolvable*), independently of our state of knowledge today or in the future.

In particular, strictly speaking, a machine learning algorithm is such a procedure, but usually the problem is only vaguely specified in advance. It produces a device (such as a neural net, decision tree or the like) to answer some set of questions that may not be exactly the ones originally intended, or produce answers that may not be correct, or fail to provide answers to some of the intended questions but not to all of them. They have to be algorithms nonetheless, executable on a physical device, either practically built (such as a conventional digital computer) or at least an ideal device that could possibly be built (such as a Turing machine.)

Much of computer science is devoted to elucidating the solvability status of APs in general, either in theory or in practice. Turing [21] demonstrated that there exist many *unsolvable* problems beyond the reach of conventional computers. For example, **MORT**(d) is algorithmically unsolvable, according to the definition above, i.e., no algorithm can ever be created or discovered by any genius that will answer the question of mortality systematically for all sets of integer matrices of size 3×3. In fact, most algorithmic problems are actually unsolvable (even ideally) and hence, *it is necessary to relax the conditions [S1]-[S3] on a solution and accept imperfect solutions that err or fail to produce exact and true answers* in a systematic fashion (systematicity is probably harder to give up since it requires intensive labor in changing devices or programs as the questions change.)

Within the realm of solvable problems, there is a further distinction that is very material for computer science and data science in particular. The definition of solution to an AP does not impose any restrictions on how long it will take for the device to produce the answers to the questions. In practice, this can become a huge issue, if the amount of resources (typically *time* or *memory*) taken by the solution device is too large, as illustrated in Table 11.4.

Table 11.4 Small and large growth rates of a running time for the brute force solution of the [SAT] problem

n	1000nlog n	100n^2	n$^{log\,n}$	...	2$^{n/3}$	2^n
20	.09 sec	.04 sec	4 sec	...	.0001 sec	1 sec.
50	.3 sec	.25 sec	1.1 hr	...	.1 sec	35 years
100	.6 sec	1 sec	220 days	...	2.7 hr	10^{10} cent.
200	1.5 sec	4 sec	125 cent.	...	3 × 10^4 cent.	?
1000	10 sec	2 min	?	...	?!	??!!

Thus, for example, to determine whether a formula containing $n = 100$ symbols (a small size in typical applications of the algorithm) is satisfiable or not would take

$$2^{100} sec/10^9 \times 3600 \times 24 \times 365 \times 100 \geq 10^8 \text{ centuries}.$$

This is obviously unacceptable since any solution device would probably take longer than the age of the universe. The trouble is, however, that this algorithmic problem **[SAT]** of satisfiability of a Boolean formula (defined above) is not just an intellectual curiosity of interest only to logicans and mathematicians. The hard fact of the matter is, *literally hundreds of problems of immediate economic importance in such diverse fields* as scheduling, combinatorial optimization (e.g., linear programming), secure secret communication (e.g., factoring and primality testing) and even national security can all be shown to be just as hard as (if not harder than) **SAT**, including **TSP**. These considerations explain, at least in part, why one of the most important problems in computing today is precisely to find new computational strategies that do overcome the complexity barrier posed by this combinatorial explosion.

As an area of computer science, *computational complexity* aims to quantify and analyze the efficiency of solutions to algorithmic problems, and hence their *inherent difficulty* if they are to be solved by any device that qualifies as a computer. To understand the basic paradigm, it is necessary to realize that the various instances of an algorithmic problem will pose various degrees of difficulty to a solution device. The situation is in contrast with other definitions of efficiency in physics and chemistry, where the efficiency of a procedure is defined by some sort of percentage measuring the amount of useful output compared to the amount of resource given. For example, in thermodynamics one measures the efficiency of a thermal engine as, say, 80% to indicate that the engine in capable of devoting eighty percent of the given amount of energy in useful work output, while spending 20% in doing its internal operations. The efficiency of a chemical reaction to produce alcohol, for example, can likewise be given as some percentage of the amount of raw materials fed to the reaction to produce it. The yield of an investment is usually given by a percentage as well.

Such a simple minded scheme cannot be used to quantify the efficiency of an algorithmic/computational problem simply because there are usually many questions to gauge the performance on; moreover, each question may take more or less work to answer. (e.g., it is much harder to figure out whether a longer number 7583 is prime integer than to decide whether 5 is prime.) An example is probably already familiar to the reader. The solutions to [**SORTING**], namely sorting algorithms, require performing a number of operations. The input size can be taken to be, for example, the number of items to be sorted, or the number of bits necessary to describe the unsorted instance file. In a comparison-based sort, it is to be expected that the bulk of the effort by a solution algorithm will be spent on performing comparisons between the various items in the input instance. Thus, a reasonable choice of resource to quantify to perform the analysis is comparisons, although it is certainly not the only one. With these measurements, the **worst-case** efficiency (or complexity) of QUICKSORT, is $O(n \log n)$. Surprisingly, the **average case** efficiency of QUICKSORT is $O(n \log n)$ as well.

In general, the measure of input size n for an arbitrary algorithmic problem in Turing computation is naturally the length on the string x describing the instance. The prime resources used in computational studies of complexity is based on the mode of operation of the solutions, which ultimately go back to Turing machines (by virtue of the Church–Turing thesis.) They are *time* (as measured by the number of elementary steps or instructions executed by the algorithms to return the answers) and *space* (the amount of memory used by the machine in the course of calculating the answer to a given instance.) Since the problem usually has many questions given by strings of variable length n, this time or memory spent becomes a function t_n or s_n of n. Since counting can become very difficult, one is only interested in the order of growth $O(t_n)$ or $O(s_n)$ of these functions for a given solution.

Once the complexity of a **particular solution** is defined, one can then make precise the complexity of an algorithmic **problem** as the lowest possible complexity achieved among all of the many possible solutions, either for time or space.

Given a bound $\{f_n\}_n$ on the amount of resource ρ available, the *complexity class* $\rho(f_n)$ consists of all algorithmic problems that admit (deterministic) solutions with complexity $O(f_n)$ units of the given resource ρ.

Thus, in the running example, **[SORTING]**, belongs to the class **TIME**$(n \log n)$ (thanks to the QUICKSORT algorithm), but it also belongs to the class **DTIME**(n^2), as does **[MINIMAL SPANNING TREE]**. **[SORTING]** does not belong, however, to the class **DTIME**(n) since QUICKSORT is an optimal general-purpose sorting algorithm.

Based on these and other considerations to become evident below, it has been proposed that the class of algorithmic problems that admit solutions that take at most a **polynomial** amount of resource $O(n^d)$ in the size n of the input, should be considered the class of *feasibly solvable* problems. The precise definition is as follows.

An algorithmic problem is *feasible* if there exists a solution whose complexity is a polynomial function in $O(n^d)$, for some integer constant $d \geq 1$. The class of *tractable/feasible* problems is the class of algorithmic problems that can be solved in polynomial time, i.e.,

$$\mathbf{P} = \bigcup_{d \geq 0} \mathbf{DTIME}\,(n^d).$$

Example 11.13 (**REC**(Primes)) is a feasible problem solvable in time $O(n^3)$ [1], a fact that took over 2300 years to establish. (The ancient Greeks were aware of the problem since they discovered there are infinitely many prime numbers.)   □

However, the feasibility of the related solvable problem [**FACTORIZATION**] of integers into prime numbers remains a challenge. Moreover, the status of most *optimization* problems of interest in data science, such as **[TSP]** or it numerous equivalents, remains unknown, i.e., some unbeknownst clever algorithm (like for primality testing) may exist that puts them in **P**. Or, perhaps they do not belong in

P at all. No one knows and some computer scientists claim humans may not have the tools to figure the answer out yet. In fact, there are outstanding prizes (like $1M dollars) for anyone with a valid proof for an answer to this vexing question for over half a century now. More details about Turing computation and complexity can be found in any textbook in theoretical computer science, e.g., [11].

Nevertheless, there is some good news concerning this class of problems (which most computer scientists believe is as good it it is going to get.) When confronted with an optimization problem (such as the shortest route in the **[TSP]** problem), *no one can stop us or any dumb machine from randomly guessing an ordering of the cities for a tour* and figure out its cost by simply adding up the costs of the edges as given for that tour. A lucky person will get a tour after a few trials that can hardly be improved with further trials. As dumb luck may have it, that may well be the optimal tour that cannot be improved at all by any other means, or at least be just *good enough* for a practical solution. Most other optimization problems exhibit similar properties, so they can be collected in a class of their own, the class **NP** (**N** stands for "nondeterministic" because their solution can only be found by guesswork, but they can be verified deterministically in polynomial time, although it may be impossible to find them out without guessing them nondeterministically.) The question of whether the two classes of algorithmic problems are identical, i.e., whether **NP** =?**P**, has remained an outstanding open problem in computer science for over half a century now, and it is recognized today as perhaps the most profound and consequential question for many sciences, data science included.

11.3.2 Machine Learning

Machine learning (ML) is commonly understood [18] as the design of algorithms/programs that improve their performance with "experience," i.e., accessing and processing data about a problem. In general, ML faces three major issues:

- *Problem representation*
  Humans have to represent the problem as a data science problem (as discussed in Sect. 11.4) as well as gathering and cleansing appropriate data (possibly including dimensionality reduction) to solve the problem. Successful data extraction often requires deep understanding of the domain of interest the problem comes from.
- *Optimization*
  ML solutions usually formulate learning tasks as an optimization problem. Typical methods used to solve these optimization problems include gradient descent, Lagrange multipliers and Maximum Likelihood.
- *Evaluation*
  Evaluating Machine learning solutions requires quantification of how well a solution works on new datasets, a nontrivial task (described in the previous subsection.) It is quite possible that the same procedure can yield very different results based on different evaluation methods.

Machine Learning (ML) can thus be better characterized as a collection of optimization procedures for learning algorithms. The most common types of techniques used to implement Machine learning solutions are described next.

Gradient Descent
This is one of the most frequently used tools in ML. The most common supervised learning algorithm, Backpropagation (described in Sect. 2.2) used in training neural networks, is a gradient descent algorithm.

Example 11.14 Gradient descent is used to find the minimum (or maximum) of a given multivariable function $f(x_1 \ldots x_n)$, i.e.,

$$\min_{x_1,\ldots,x_n} f(x_1, \ldots x_d)$$

For such a function, the direction (of the many inputs x_i) of fastest change from a point $\mathbf{x}$ (the steepest slope in the container space $(d + 1)$D space of the graph of the function) is indicated by a vector called the *gradient* whose components are the partial derivatives with respect to each of the independent variables in the function evaluated at the point $\mathbf{x}$, i.e.,

$$\nabla f(x_1, \ldots x_n) = \left[\frac{\partial f}{\partial x_1}(\mathbf{x}), \ldots, \frac{\partial f}{\partial x_n}(\mathbf{x}) \right]$$

Naturally, this direction changes with the specific point $\mathbf{x}$. To minimize f, one can start ($t = 0$) at a nearby point $\mathbf{x}^0$ and take a small step in the direction that is opposite to the gradient direction, i.e., that of $-\nabla f(\mathbf{x}^0)$. To update the candidate optimal location $\mathbf{x}^{(t)}$ for a minimum, the x_i in iteration $t > 0$ is given by

$$x_i^{(t)} = x_i^{(t-1)} - \epsilon \frac{\partial f}{\partial x_i}(\mathbf{x}^{(t-1)})$$

where ϵ is a small positive constant. A choice of a very small ϵ may take a long time to converge, while a large ϵ may jump over the optimal location for a minimum value of f. For convex functions (with a single global minimum), gradient descent is guaranteed to find the location where f attains its minimum value, the optimal solution. □

Max-Likelihood Estimation
Max-Likelihood Learning, also called Max-Likelihood Estimation (MLE), is a general method for learning probabilistic models. MLE learns the parameters of a distribution from observations. Given independent and identically distributed (i.i.d) observations, $x_1 \ldots x_n$, MLE aims to optimize the function

$$\max_{\theta_1,\theta_2,\ldots,\theta_k} P(x_1 \ldots x_n \mid \theta_1, \theta_2, \ldots, \theta_k)$$

where $\theta_1, \theta_2, \ldots, \theta_k$ represent the parameters of the distribution. To estimate these parameters, one can use a gradient ascent procedure (analogous to the gradient descent procedure above, moving in the direction of $\nabla P(\Theta^0)$.)

Lagrange Multipliers
This method is common in multivariable calculus to optimize functions with several variables, usually when there are constraints on the variables.

Example 11.15 Support Vector Machines (SVMs) employ this technique of optimization. In order to maximize a function $f(x_1, x_2)$ subject to a constraint between x_1 and x_2 that $g(x_1, x_2) \geq 0$, one considers a multiplier λ (called a *Lagrange multiplier*) and reformulate the problem as an optimization problem with an extra variable λ to maximize

$$L(x_1, x_2, \lambda) = f(x_1, x_2) + \lambda g(x_1, x_2).$$

To solve the optimization problem in terms of the variables x_1, x_2 and λ, the Karush–Kuhn–Tucker conditions (KKT conditions) specify that one can solve the optimization problem in terms of three constraints.

$$g(x_1, x_2) \geq 0$$

$$\lambda \geq 0$$

$$\lambda g(x_1, x_2) = 0$$

Thus, whenever $\lambda \neq 0$, $g(x_1, x_2) = 0$. In SVMs, points corresponding to nonzero Lagrange multipliers are referred to as support vectors. Intuitively, they represent the boundary of the function. The stationary points at $\lambda \neq 0$ turn out to be the points at the boundary of the function, i.e., where $g(x_1, x_2) = 0$.   □

Gaussian Learning
Gaussian distributions (described in Appendix 11.1) are widely applied in unsupervised learning, including Gaussian Mixture Models and autoencoders. For instance, a single variable Gaussian distribution is defined by 2 parameters, the mean μ ($-\infty < \mu < \infty$) and variance ($\sigma^2 \geq 0$). The distribution is defined as follows.

$$N(X \mid \mu, \sigma^2) = \frac{1}{(2\pi\sigma^2)^{1/2}} \exp\left(-\frac{1}{2\sigma^2}(x - \mu)^2\right)$$

$$E(X) = \mu,$$

$$Var(X) = \sigma^2,$$

$$Mode(X) = \mu.$$

In a multivariable situation of dD vectors ($d > 1$), the Gaussian parameters include a d-dimensional mean vector and $d \times d$ covariance matrix that is both symmetric and positive-definite (described in Appendices 11.1 and 11.2.)

11.4   Typical Data Science Problems

Data science problems are defined as computational algorithmic problems and are assigned a mnemonic name in brackets (e.g., [**IrisC**]) used to refer to the problem elsewhere. The last character indicates the type of problem they are (**Classification, Prediction** or **Clustering**.) The population of inputs is described in the INSTANCEs of the problem, together with an example of such. The problems are usually solved using certain samples described in the next Appendix 11.5, where more details about the population can be gathered.

[AstroP]   PLANETARY BODY LOCATION

INSTANCE: A celestial body x in our solar system (e.g., Mars or Saturn), a location on Earth (latitude, longitude) and a time stamp t

QUESTION: What is the location (azimuth and elevation) in the sky where x will be found from the given location at time t?

[IrisC]   IRIS FLOWER CLASSIFICATION ({ Setosa, Versicolor, Virginica })

INSTANCE: A feature vector
$x = $ (sepal length, sepal width, petal length, petal width) (in cms) describing an iris flower, e.g., $(4.6, 3.1, 1.5, 0.2)$

QUESTION: Which kind of flower is x?

[NoisC]   NOISY CLASSIFICATION

INSTANCE: A synthetic dataset D where some (derived) features have been obtained by certain combinations (linear or statistical) from the remaining hidden (primitive) features

QUESTION: Which are the derived and the primitive features in D?

[CharRC]   CHARACTER RECOGNITION ({0, 1, 2, 3, 4, 5, 6, 7, 8, 9})

INSTANCE: A 2D picture of a handwritten decimal numerical digit x

QUESTION: Which digit is it?

[MalC]   MALWARE CLASSIFICATION (Π)

INSTANCE: A piece of malware x (program)

QUESTION: Which category c in partition Π does x belong in?

The categories in the partition Π could be those listed in Table 11.13 in Appendix 11.4.

[BodyFP]   HUMAN BODY FAT ESTIMATION

INSTANCE: A feature vector with a person x's body measurements including weight, height, age and others[1]

QUESTION: What is x's body fat percent of the total weight?

The full set of features is listed in Table 11.9 in Sect. 11.4.

[BioTC]   BIOTAXONOMIC CLASSIFICATION (T)

INSTANCE: A (long) DNA sequence (over the alphabet $\{a, c, g, t\}$) from a living organism x

QUESTION: What species in taxon T does x belong in?

The features representing an organism could be, for example, mitochondrial genes COI, COII, COIII, and CytB from the organism's genome. Other choices give rise to different problems, as shown in the corresponding dataset in Sect. 11.5. A most complex problem will arise if the full genome is used to represent an instance of a biological organism.

[PhenoP]   PHENOTYPIC PREDICTION (T, F)

INSTANCE: A (long) DNA sequence x (over the alphabet $\{a, c, g, t\}$) from a living organism in T?

QUESTION: What's the quantitative measurement of x's phenotypic feature F when fully grown?

Examples of phenotypic features are the area of the cephalic apotome (head top), its peculiar spot pattern, the body color of its thorax/abdomen, or the area of the postgenal cleft (mandible) in a black fly larva; Or, it could be the rosette dry mass (leaf weight), or the life span of a specimen of A. *thaliana* when fully grown (end of the reproductive cycle to adulthood.)

[RossP]   ROSETTE AREA OF A PLANT (T)

INSTANCE: A (long) DNA sequence x (over the alphabet $\{a, c, g, t\}$) from a plant in T?

QUESTION: What's the area of x's rosette when fully grown?

[LocP]   PROVENANCE OF A PLANT (T)

INSTANCE: A (long) DNA sequence x (over the alphabet $\{a, c, g, t\}$) describing a plant in T?

QUESTION: Where on Earth (latitude, longitude) was x grown?

[CWD]   CODEWORD DESIGN

INSTANCE: A positive integer m and a threshold $\tau > 0$

QUESTION: What is a largest set B of single DNA strands of length m that do not cross hybridize to themselves or to their complements (nxh set) under stringency τ according to the h-distance , i.e., $|xy| > \tau$ for all $x, y \in B$?

11.5   A Sample of Common and Big Datasets

This section summarizes datasets that are used throughout the book to illustrate methods and solutions for a variety of dimensionality reduction methods and problem solving, or that could be used in other applications. (They are organized by domain area and are listed in alphabetic order.)

Astronomy

This dataset is perhaps one the oldest and most comprehensive and accurate set of observations (for his time) of the skies with the naked eye, compiled over a period of 40+ years by the Danish astronomer Tycho Brahe and his assistants with the naked eye (telescopes were not invented yet) in the late 1500s. His student Johannes Kepler used the data to synthesize our current model of the solar system that reproduces the data based in the knowledge expressed in the three well-known physical Kepler's laws (Table 11.5).

Biology

The Iris Flower dataset was used by Fisher to test his ideas about Discriminant Analysis [7]. It is perhaps one of the oldest and well-known dataset in the statistics, pattern recognition and data science literature. It contains 3 classes of 50 instances each, where each class refers to a type of iris plant, as illustrated in Tables 11.6 and 11.7.

The Body Fat dataset contains estimates of the percentage of body fat for 252 men determined by underwater weighing and various body circumference measurements. (Accurate measurement of body fat is inconvenient/costly and it is desirable to have alternative methods of estimating body fat based on more easily available measurements.) The set was assembled and published by [12] (Tables 11.8, 11.9, and 11.10).

Table 11.5 Tycho Brahe's dataset

Population	Celestial bodies
Dimensionality (features)	Hundreds
Class distribution	N/A
Size (Nr of data points)	Tens of thousands
Sample data point	Positions of Mars and comets in the celestial sky
Origin	[3]

Table 11.6 The Iris flower dataset

Population	Iris flowers
Dimensionality (features)	4
Class distribution	50 50 50
Size (nr of data points)	150
Sample data point	(4.6, 3.1, 1.5, 0.2)
Origin	[7]

Table 11.7 Iris dataset (lengths are given in *cm*)

ID	Sepal length	Sepal width	Petal length	Petal width	Label (species)
1	5.1	3.5	1.4	0.2	Iris-setosa
2	4.9	3.0	1.4	0.2	Iris-setosa
...	...	...	...	...	...
50	5.0	3.3	1.4	0.2	Iris-setosa
51	7.0	3.2	4.7	1.4	Iris-versicolor
52	6.4	3.2	4.5	1.5	Iris-versicolor
...	...	...	...	...	...
100	5.7	2.8	4.1	1.3	Iris-versicolor
101	6.3	3.3	6.0	2.5	Iris-virginica
102	5.8	2.7	5.1	1.9	Iris-virginica
...	...	...	...	...	...
150	5.9	3.0	5.1	1.8	Iris-virginica

Table 11.8 Human body fat dataset

Population	Humans
Dimensionality (features)	4
Size (Nr of data points)	252
Sample Data Point	$(154.25, 67.25, \cdots, 17.10, 23, 12.30)$
Origin	[12]

Table 11.9 Examples of 16 features in the body fat dataset of 252 men (columns.) (If unspecified, circumferences are given in *cm*)

	Person IDs					
Feature	1	2	3	4	...	252
Weight (lbs)	154.25	173.25	154.00	184.75	...	207.5
Height (in)	67.75	72.25	66.25	72.25	...	70
BMI($= \frac{W}{H}$ in Kg/m)	23.70	23.40	24.70	24.90	...	29.80
Neck	36.20	38.50	34.00	37.40	...	40.8
Chest	93.10	93.60	95.80	101.80	...	112.4
Abdomen	85.20	83.00	87.90	86.40	...	108.5
Hip	94.5	98.70	99.20	101.20	...	107.1
Thigh	59.00	58.70	59.60	60.10	...	59.3
Knee	37.30	37.30	38.90	37.30	...	42.2
Ankle	21.90	23.40	24.00	22.80	...	24.6
Bicep	32.00	30.50	28.80	32.40	...	33.7
Forearm	27.40	28.90	25.20	29.40	...	30
Wrist	17.10	18.20	16.60	18.20	...	20.9
Age (years)	23	22	22	26	...	74
BodyFat (lbs)	12.30	6.10	25.30	10.40	...	31.9

Table 11.10 The biotaxonomy dataset

Population	Organisms in 21 species across the biome
Dimensionality (features)	500-10,000
Class distribution	(Table 11.11)
Size (data points)	249
Typical data point	(COI, COII, COIII, CytB) (DNA sequences)
Origin	[16]

Table 11.11 Organisms in the sample data for species classification of a biological taxon. The 21 classes/labels in the partition are in the first column

Label	T: Genus species	Common name	Count
1	Apis mellifera	Western honey bee	4
2	Arabidopsis thaliana	Thale cress	5
3	Bacillus subtilis	Hay/grass bacillus	18
4	Branchiostoma floridae	Florida lancelet	18
5	Caenorhabditis elegans	Round worm	6
6	Canis lupus	Wolf	18
7	Cavia porcellus	Pork	4
8	Danio rerio	Zebra fish	9
9	Drosophila melanogaster	Fruit fly	18
10	Gallus gallus	Red junglefowl	18
11	Heterocephalus glaber	Naked mole rat	3
12	Homo sapiens	Human	18
13	Macaca mulatta	Rhesus macaque	8
14	Mus musculus	House mouse	18
15	Neurospora crassa	Red bread mold	5
16	Oryza sativa	Asian rice	12
17	Pseudomonas fluorescens	Infectious bacterium	6
18	Rattus norvegicus	Brown rat	18
19	Rickettsia rickettsii	Tick-born bacterium	18
20	Saccharomyces cerevisiae	Yeast	18
21	Zea mays	Corn/maize	7
			249

The Biotaxonomy dataset was obtained from 249 organisms collected from genetic repositories (e.g., GenBank at www.ncbi.nlm.nih.gov/genbank/) and is illustrated in Tables 11.11 and 11.12.

Cybersecurity

The Malware Classification Challenge (*arxiv.org/abs/1802.10135*) [2, 19] dataset was released in 2015 and is publicly available through *kaggle.com*. An available dataset consists of a set of 10,868 known malware files representing a mix of nine (9) different malware families, as summarized in Table 11.13, e.g., "ramnit." *Code.exe* is a computer executable binary file obtained from compiling the source malware

Table 11.12 Microsoft's Malware classification dataset

Population	Malware
Dimensionality (features)	1804
Class distribution	1541 2478 2942 475 42 751 398 1228 1013
Size (data points)	10,868
Sample data point	(*Code.exe*, . . ., Vundo)
Origin	*kaggle.com*

Table 11.13 Microsoft's Malware classification problem. The labels for the classes in the partition are in the first column

Labels	Class name	Type of Malware	*Count*
1	Ramnit	Worm	1541
2	Lollipop	Adware	2478
3	Kelihos_ver3	Backdoor	2942
4	Vundo	Trojan	475
5	Simda	Backdoor	42
6	Tracur	TrojanDownloader	751
7	Kelihos_ver1	Backdoor	398
8	Obfuscator.ACY	Any kind of obfuscated malware	1228
9	Gatak	Backdoor	1013
			___
			10, 868

(headers removed.) Each datapoint contains both the raw binary content of the malware file as well as metadata information extracted using the IDA disassembler tool. These features include source code snippets, assembly command frequencies, registers used and their frequency, data section of the code, key words in the disassembly code, as well as a number of other extracted features (e.g., entropies of the various features.) The classification challenge was to correctly assign every piece of malware to one of the nine (9) categories.

Finance

The Adult dataset was extracted from the 1944 Census Bureau database found at the source by Barry Becker to predict whether a person makes over 50, 000 a year (Tables 11.14 and 11.15).

The Audit dataset was used to support research by the Ministry of Electronics and Information Technology (MEITY) of the government of India. The goal was to help auditors classify whether a firm is fraudulent or not on the basis of historical and current risk factors (Table 11.16).

The Bank Marketing dataset was designed by a Portuguese banking institution to direct marketing campaigns on the phone to decide whether a client will subscribe (yes/no) to a term deposit (Table 11.17).

Table 11.14 Adult dataset

Population	Census income
Dimensionality (features)	15
Class distribution	23.93% 76.07%
Size (nr of data points)	48,842
Sample data point	(53, Confidential, 234, 721, 11th, 7, Married-civ-spouse, Handlers -cleaners, Husband, Black, Male, 0, 0, 40, United-States, $\leq 50K$)
Origin	*www.census.gov/ftp/pub/DES/www/welcome.html*

Table 11.15 Audit dataset

Population	Fraudulent firms
Dimensionality (features)	27
Class distribution	114 77 82 70 47 95 1 4 5 3 1 41 37 200
Size (nr of data points)	776
Sample data point	(3.89, 6, 0, 0.2, 0, 10.8, 0.6, 6.48, 10.8, 6, 0.6, 3.6, 11.75, 0.6, 7.05, 2, 0.2, 0.4, 0, 0.2, 0, 4.4, 17.53, 0.4, 0.5, 3.506, 1)
Origin	*archive.ics.uci.edu/ml/machine-learning-databases/00475/*

Table 11.16 Bank marketing dataset

Population	Bank clients
Dimensionality (features)	17
Class distribution	N/A
Size (nr of data points)	45,211
Sample data point	(47, blue-collar, married, N/A, no, 1506, yes, no, N/A, 5, may, 92, 1, −1, 0, N/A, no)
Origin	*archive.ics.uci.edu/ml/machine-learning-databases/00222/*

Table 11.17 Taiwanese bankruptcy dataset

Population	Customer default payers
Dimensionality (features)	24
Class distribution	Unknown
Size (nr of data points)	30,001
Sample data point	(50, 000, 2, 2, 1, 37, 0, 0, 0, 0, 0, 0, 46, 990, 48, 233, 49, 291, 49, 291, 28, 314, 28, 959, 29, 547, 2000, 2019, 1200, 1100, 1069, 1000, 0)
Origin	*archive.ics.uci.edu/ml/machine-learning-databases/00350/*

The Taiwanese Bankruptcy dataset was used to evaluate the predictive accuracy of six data mining probability models of default based on labeled customer data (Yes/No) on default payments in Taiwan.

Image Processing

The MNIST dataset [15] consists of gray scale images of handwritten digits. The digits were handwritten by members of the US Census Bureau and high school students. The digit in each image is size-normalized and centered. The dataset is fairly large. The MNIST dataset is a popular dataset and a benchmark in the image processing, machine learning and neural network literature. One of its major real world applications was in the postal service to automatically scan and infer handwritten zip codes. The pixels are binary images, i.e., the pixels are either on or off (0 or 1.) (Tables 11.18 and 11.19).

The ImageNet dataset [14] is the largest hand-labeled image dataset currently available. The latest dataset was published for the ImageNet Large Scale Visual Recognition Challenge 2017 (ILSVRC17) for the problem of object detection in and classification of images, i.e., to identify and find the location of objects familiar to humans in images in a bounding box. The dataset was inspired to create a standard benchmark (North Star) for computer vision research, is organized according to the WordNet ontology (nouns only) and has been expanded to include 1000 categories for object localization and 30 fully labeled categorized for object detection in videos. It is also publicly available through *kaggle.com*.

Leisure

The Old Faithful Geyser dataset was singled out by Azzalin and Bowman [4] and is available as a standard set in the package R or at *www.stat.cmu.edu/~larry/all-of-statistics/=data/faithful.dat*. Eruptions have been clustered into 3 categories (Short, Medium, Long) (Tables 11.20 and 11.21).

Table 11.18 MNIST dataset

Population	28 × 28D handwritten images of decimal digits
Dimensionality (features)	784
Class distribution	0: 5923, 1: 6742, 2: 5958, 3: 6131, 4: 5842, 5: 5421, 6: 5918, 7: 6265, 8: 5851, 9: 5949
Size (nr of data points)	60,000
Sample data point(s)	(Figs. 2.2 and 5.6)
Origin	[15]

Table 11.19 ImageNet dataset

Population	Images of objects familiar to humans
Dimensionality (features)	181,503
Class distribution	200 object categories;
	456,567 images of 478,807 objects in a training set;
	40,152 images in a testing set;
	20,121 images of 55,502 objects in a validation set
Size (nr of data points)	516,840
Sample data point	N/A
Origin	*image-net.org*

Table 11.20 Old faithful Geyser dataset

Population	Eruptions of old faithful (Yellowstone national park)
Dimensionality (features)	2
Class distribution	N/A
Size (nr of data points)	272
Sample data point	(3.600, 79) =(Duration, wait time to next eruption) (mins)
Origin	[7]

Table 11.21 Netflix movie rating dataset

Population	Netflix movies
Dimensionality (features)	4
Class distribution	N/A
Size (nr of data points)	2,817,131
Sample data point	Quadruplets <user, movie, date of rating, rating>
Origin	*www.kaggle.com/laowingkin/netflix-movie-recommendation*

Table 11.22 The Ames housing dataset

Population	Residential properties in Ames, Iowa
Dimensionality (features)	82
Class distribution	N/A
Size (nr of data points)	2930
Sample data point	82D vectors describing property/houses
Origin	Assessor's office,*jse.amstat.org/v19n3/decock/DataDocumentation.txt*

Th Movie Rating dataset was provided by Netflix to support an open competition to develop a movie recommendation system in 2006. The dataset is publicly available through *kaggle.com*. The movie rating files contain ratings (1 to 5 *s) from 480,189 randomly chosen, anonymous Netflix customers (not associated with Netflix nor in certain blocked countries) over 17,770 movie titles. The data were collected between October, 1998 and December, 2005 and reflect the distribution of all ratings received during this period. The average user rated 2000 movies and each movie was rated by over 5000 users on the average, with wide variation (some movies received only 3 ratings and one user rated over 17,000 movies.) Although no user information is provided, the dataset has been has been criticized by privacy advocates for leaking customer information, although it was supposed to have been constructed to preserve customer privacy (Table 11.22).

The Ames Housing dataset contains information from the Ames Assessor's Office about the assessed values for individual residential properties sold in Ames, Iowa from 2006 to 2010 (Fig. 11.1). The data has 82 columns which include 23 nominal, 23 ordinal, 14 discrete, and 20 continuous variables (and 2 additional observation identifiers).

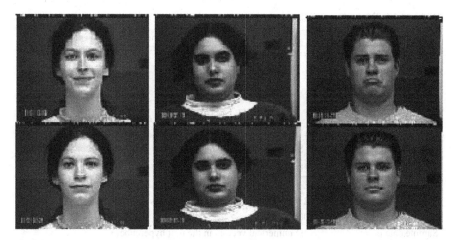

Fig. 11.1 Facial expressions (positive: *left*; neutral: *middle*; negative: *right)* by trained actors in the Cohn-Kanada dataset

Table 11.23 The Cohn-Kanade dataset

Population	Humans
Dimensionality (features)	64
Size (Nr of data points)	500 from 100 actors
Sample Data Point	(Fig. 11.1)
Origin	[13]

Psychology

The Cohn-Kanade dataset consists of approximately 500 image sequences from 100 subjects ranging in age from 18 to 30 years (65% are female; 15% are African–American and 3% are Asian or Latino.) Each begins from a neutral or nearly neutral face. For each, an experimenter described and modeled the target display. Six are based on descriptions of prototypic emotions (i.e., joy, surprise, anger, fear, disgust, and sadness.) These six tasks and mouth opening in the absence of other action units were annotated by certified FACS coders [6]. Seventeen percent (17%) of the data were comparison annotated. Inter-observer agreement was quantified with the kappa coefficient, which is the proportion of agreement above what would be expected to occur by chance [22] [8], with an average observer agreement of 0.86. Image sequences from neutral to target display were digitized into 640 × 480 or 640 × 490 pixel arrays with 8-bit precision for grayscale values. (Further details can be found at *://www.ri.cmu.edu/project/cohn-kanade-au-coded-facial-expressiondatabase/* (Table 11.23).

Table 11.24 US DOE campus safety and security dataset

Population	Title IV College Campuses in the US
Dimensionality (features)	Hundreds
Class distribution	(Variable by year)
Size (data points)	Nearly 11,000 campuses
Sample data point	N/A
Origin	US Department of Education at *ope.ed.gov/campussafety*

Table 11.25 Synthetic dataset SYN12

Population	12D unit cube in Euclidean space
Dimensionality (features)	12
Class distribution	12 (6 primitive and 5 derived features, plus one label)
Size (nr of data points)	1000
Sample data point	12D numerical vectors in the unit hypercube
Origin	[16]

Table 11.26 Synthetic dataset SYN23

Population	23D unit cube in Euclidean space
Dimensionality (features)	12
Class distribution	23 (12 primitive and 10 derived features, plus one label)
Size (nr of data points)	500
Sample data point	23D numerical vectors in the unit hypercube
Origin	[16]

Safety in College Campuses

The Department of Education (DOE) of the United States collects and dissemi-
nates the yearly crime statistics for all title IV colleges and universities in the
United States. Major crime categories include arrests, criminal offenses, disci-
plinary actions, fire statistics, hate crimes, VAWA offenses, and sexual misconduct
(Table 11.24).

Synthetic Data

These synthetic datasets are fully described in the reference above [16] where
the sets originated. They afford perfect prior knowledge of *feature dependencies*
that can remain hidden from the DR methods. The effectiveness of DR methods
can then be assessed by how well they discover these hidden but most relevant
and independent features, from a full set that includes other confounding features
derived from the few primitive ones (Tables 11.25 and 11.26).

The primitive features were designed using the method (API) described in [10]
and publicly available in Python at *sklearn.datasets.make_classification*. The second data set was generated likewise but halving the number of all parameters
involved in generating features in the second dataset SYN23, so just the second and
largest set is described. The dataset was generated in two phases. First, the sklearn
API was used to generate the primitive features in a 12D hypercube with sides of

length $1.6 = 2 * 0.8$ and assign about the same number of points to each of the eight classes. Each class is composed of several Gaussian clusters located near corners of the 12D hypercube. For each class, the informative features are drawn independently from the standard normal distribution $N(0, 1)$. Four more features X_{13}, X_{14}, X_{15} and X_{16} were generated as linear combinations (with random coefficients) of some primitive features, as provided in the API. In the second phase, 6 more predictors were generated as follows. Features X_{17} and X_{18} were drawn as a repeat of two randomly selected primitive features/columns. X_{19} was generated as the sum of squares of two features selected randomly as

$$X_{19} = X_i^2 + X_j^2 \text{ for some fixed } 0 \le i, j \le 12.$$

X_{20} was obtained from the values of a linear regression model fitted using two randomly selected features X_i and X_j as predictors with the *Label* as response. The value for X_{20} was the deviation between the pair of values in *Label* and the regression prediction. X_{21} was obtained the same way but with X_i^2 and X_j as predictors from another random selection of predictors X_i and X_j. Finally, X_{22} was generated as the outcome of the natural *logarithm* of a randomly selected but uniform predictor X_i (a transformation that does not change entropy.) Again, different types of predictors were used as described above.

In general, synthetic data has a number of uses in data science. First, as illustrated with datasets SYN12 and SYN23, they may increase prior knowledge about the data and allow for much greater control on the part of the analyst to test, validate and assess data analyses. Second, they may alleviate a number of concerns making data available to the analyst, including *privacy* (e.g., with health records), *anonymity* (legal compliance with, e.g., HIPPA regulations), *intellectual property*, *value* (data is nowadays considered to be the most valuable asset in an organization) and *accessibility*. considered to be probably the major roadblick to data analytics.) Where any of these issues is a concern, real datasets can be cleansed and/or transformed into roughly equivalent synthetic datasets that preserve the critical trends and structure of the real data while being completely "fake," thus affording results of the analyses that nonetheless apply to the original data and problems. Third, using the DR techniques discussed in Chap. 8, they can be converted into much more informative datasets while being smaller in size and easier to crunch and analyze. This can be very useful in large datasets where a pilot feasibility study would inform a decision whether the results will be worth a large and costly effort and even whether a project is feasible. It is therefore not surprising that most data analytics platforms (described in Appendix 11.6) offer facilities to generate synthetic data for specific problems and/or solutions (e.g., classification, clustering, regression.)

Text Processing

WordNet is a lexical database of semantic relations between abstract language concepts (so-called *synsets* of cognitive synonyms) in a given natural language (more than 200 now), including synonyms, hyponyms, and meronyms. WordNet has been used for many purposes in information systems, including word-sense disambiguation, information retrieval, automatic text classification and text summarization, machine translation and even automatic crossword puzzle generation. WordNet was originally a lexical database for the English language created by researchers at Princeton led by George Miller as part of the NLTK corpus (Table 11.27). It is regularly updated by the Princeton WordNet group, whose aim is to ensure that there is an up-to-date version of high quality available for the English language. It is available at *https://CRAN.R-project.org/package=wordnet*.

Other Data Collections

Many other datasets are constantly being generated or updated, so it is impossible to account for them all. A selection of other sources that may come in handy currently are shown in Table 11.28, particularly for big data sets.

Table 11.27 WordNet

Population	Distinct synsets (representing abstract semantic concepts)
Dimensionality (features)	Labeled edges between synsets indicating lexical relationships (like antonym, meronym, part-of-speech) in the WordNet tree
Class distribution	N/A
Size (nr of data points)	117,000
Sample data point(s)	(leg, meronym, chair), (arm-chair, hyponym, chair), (love, antonym, hate)
Origin	[17]

Table 11.28 Other available big dataset collections

Name	Source	Description
Digital sky survey	*sdss.org*	The Sloan Digital Sky Survey has collected and organized outer space astronomical observations (order of terabytes daily) that has produced the most detailed 3D maps of the Universe ever made
EARTHDATA	*https://earthdata.nasa.gov/*	NASA's Common Metadata Repository (CMR) is a high-performance, high quality, continuously evolving metadata system that catalogs all data and service metadata records for NASA's Earth Observing System Data and Information System (EOSDIS)
Marine Geoscience Data System (MGDS)	*www.re3data.org/ repository/r3d100010273*	Data repository that provides free public access to marine geophysical data products and datasets relevant to the formation and evolution of the seafloor and sub-seafloor

(continued)

Table 11.28 (continued)

Name	Source	Description
GenBank (NIH)	*www.ncbi.nlm.nih.gov/ genbank/*	US National Institute of Health genetic sequence data repository, an annotated collection of all publicly available DNA sequences
IEEE dataPort	*ieee-dataport.org/datasets*	Over 3000 datasets in 25+ categories of topics, including biomedicine, COVID-19, ecology, environment, power and energy, social science and transportation
NOAA Data Catalog	*https://data.noaa.gov/ dataset/*	US National Oceanic and Atmospheric Administration's datasets, including oceans, currents, glaciers, temperature and barometric pressure
UCI Machine Learning Repository	*https://archive-beta.ics.uci. edu/*	600 datasets about a variety of topics, from flowers, wine, diabetes and heart disease, to auction verification, student academic success and ImageNet, to air quality, energy efficiency, the ozone. The datasets are organized by various criteria (e.g., subject, problems/tasks) and properly documented (original design and problems.) There is a facilitate to donate datasets
Visual Datasets	*https://public.roboflow. com/*	About 40 datasets consisting of images on topics including mushrooms, aerial maritime drones, wildfire smoke, pistols, and self-driving cars

11.6   Computing Platforms

This section provides summaries of computing platforms for data science and basic tips to get started with the software. More detailed tutorials can be found readily available online.

11.6.1   The Environment R

R is a popular interactive environment for computational programming and statistical analysis. It is also considered to be a programming language. R can be traced back to S, a language and package developed at Bell Laboratories of AT&T by John Chambers and others (in 1976) for statistical analysis that was originally free but eventually became a commercial package, S-PLUS. In 1991, R was created by

Ross Ihaka and Robert Gentleman in the Department of Statistics at the University of Auckland and it become available to the public in 1993. Since S-Plus and R are implementations of S, both products can execute the same functions and most common statistical functions without modification.

R is freely available for download from a comprehensive R archive network, CRAN at *cloud.r-project.org*. CRAN is composed of a set of mirror servers distributed around the world. There is an extension, RStudio that offers a user interface in an integrated development environment that can be downloaded from *www.rstudio.com/download*.

R has several software packages (few system supplied and many user-contributed) available to implement various supervised and unsupervised statistical and machine learning algorithms and evaluation methods. A list of the important software libraries in R is shown in Table 11.29. The packages can be downloaded

Table 11.29 Popular software packages in R

Package::FunName()	Description
`stats::lm()`	fits linear models
`stats::glm()`	fits generalized linear models (GLMs)
`glmnet::glmnet()`	fits a GLM with LASSO or ElasticNet regularization
`base::svd()`	returns the singular value decomposition of a rectangular matrix
`stat::princomp()`	returns the principal components of the given data matrix in an object of class princomp (via spectral decomposition)
`stat::prcomp()`	returns the principal components of the given data matrix in an object of class prcomp (via singular value decomposition-SVD)
`fastICA::fastICA()`	performs Independent Component Analysis (ICA)
`kernlab::kpac()`	performs kernel PCA (KPCA)
`dimRed::`	various dimensionality reduction methods implemented in R (e.g., PCA, KPCA and nonlinear methods **MDS**, **ISOMAP** and *t*-**SNE**)
`e1071::svm()`	fits a support vector machine-SVM for general regression and classification
`randomForest::randomForest()`	fits a random forests for classification and regression
`mlogit::mlogit()`	fits a multinomial logit model for estimation and classification (possibly with alternative-specific and/or individual-specific variables)
`tree::tree()`	fits a classification or regression tree
`class::knn()`	fits a k-Nearest Neighbor classifier
`MASS::lda()`	performs a linear discriminant analysis
`MASS::qda()`	performs a quadratic discriminant analysis
`infotheo::entropy()`	returns the values of the Shannon entropy(ies) of the given data
`infotheo::condentropy()`	returns the conditional entropy of the given random variables

from *cloud.r-project.org*, along with tutorials and demonstrations for the packages. Some handy commands to get a quick start with the methods discussed in this book are shown in Table 11.30.

Table 11.30 Handy commands in R for dimensionality reduction in data science

Command	Action	Notes
`help`	provide documentation/help with any topic	`help(help)` shows more help with how to use `help`
`y<-eigen(A)`	assigns dataframe `y` the spectral decomposition of matrix A	`y$val` are the eigenvalues; `y$vectors` are the eigenvectors
`y<-svd(A)`	assigns dataframe `y` the SVD of matrix A	`y$d` is the vector of singular values of A
		`y$u` is a matrix with the left singular vectors in the columns
		`y$v` is a matrix with the right singular vectors in the columns
`PC <- prcomp(X)`	assigns dataframe `PC` the principal components of data X	`PC$rotation` contains the rotated data (in the PC axes)
`PC <- prcomp(X, retx = TRUE, center = TRUE, scale. = FALSE tol = NULL, rank = NULL, ....)`	`retx$d = TRUE` iff rotated values are returned; `center/scale=TRUE` iff the data X is centered/scaled to Z-scores (advisable in general, but *scale* cannot be used if there are zero or constant variables); `subset` is an optional vector used to select rows from data X;	defaults for `prcomp`; components are omitted if their standard deviations are less than or equal to `tol` times the standard deviation of the first component; `rank` can be used instead to select only the max rank of X number of components;
`PC <- princomp(X)`	similar to `prcomp` for PCs, but `princomp` is a generic function with "formula" and "default" methods	`PC$rotation` contains the rotated data (in the PCs axes)
`predict(object, newdata, ...)`	predicts values for the `newdata` using the SVD of `object`	
`entropy(X, method="emp")`	computes the entropy using estimator emp	X must be discrete; the option `equalfreq` can be used to discretize where necessary)
`condentropy(X, Y=NULL, method="emp")`	computes the conditional entropy $H(X\|Y)$ using estimator method emp	as with `entropy`

11.6.2   Python Environments

A general list of software packages useful in data science is shown in Table 11.31.

The *scikit-learn* (sklearn) module in Python is a high-level module that includes all libraries required for pre-processing data, implementing supervised and unsupervised machine learning algorithms and computing evaluation metrics. A list of the important libraries within sklearn is shown in Table 11.32. The packages can be downloaded from *scikit-learn.org*, where tutorials and demonstrations how to use the packages are also available.

Tables 11.33 and 11.34 show software libraries and commonly used commands in Python (sklearn) for dimensionality reduction and supervised/unsupervised Machine learning methods.

Table 11.31 Data science software in python

Name	Description	Link
Scikit-Learn	Python module for data science	*scikit-learn.org*
WEKA	Machine Learning software in Java	*www.cs.waikato.ac.nz/ml/weka/*
R	Several packages related to data science	*cran.r-project.org/*
Matlab	Statistics and Machine Learning Toolbox	*www.mathworks.com/products/statistics.html*
TensorFlow	Deep Learning Libraries	*www.tensorflow.org/*
PyTorch	Deep Learning Libraries	*pytorch.org/*
Keras	Python Deep Learning APIs	*keras.io/*
Spark MLlib	Machine Learning for Big Data	*spark.apache.org/mllib/*
Google Vertex AI	Machine Learning on the Cloud	*cloud.google.com*

Table 11.32 Scikit-Learn Software packages in python

Name	Description
sklearn.preprocessing	Feature scaling, imputation of missing values
sklearn.decomposition	Dimensionality reduction (PCA, KernelPCA, Nonnegative Matrix Factorization, Singular Value Decomposition)
sklearn.manifold	Nonlinear Dimensionality reduction (**ISOMAP**, *t*-**SNE** and **MDS**)
sklearn.feature_selection.RFE	Wrapper methods for feature selection
sklearn.model_selection	Model selection using cross-validation and other approaches
sklearn.model_selection.metrics	Metrics for supervised learning (e.g., precision, recall, and F1-score) and unsupervised learning (e.g., silhouette score)
sklearn.neighbors	k-Nearest Neighbors
sklearn.naive_bayes	Naive Bayes Classifier
sklearn.tree	Decision Trees
sklearn.ensemble	Bagging, Boosting and other ensemble methods
sklearn.neural_network	Neural Networks
sklearn.svm	Support Vector Machines
sklearn.cluster	Clustering algorithms
sklearn.mixture	Gaussian Mixture Models

Table 11.33 Handy commands in sklearn for dimensionality reduction in data science

Command	Action	Notes
`feature_selection.RFE` `(estimator,` `n_features_to_select)`	Creates an RFE feature selection object	The base classifier `estimator` selects `n_features_to_select`
`manifold.Isomap(n_neighbors,` `n_components)`	Creates an object used to perform **ISOMAP** based dimensionality reduction	uses `n_neighbors` for each point and `n_components` coordinates for the manifold
`manifold.LocallyLinear` `Embedding(n_neighbors,` `n_components)`	Creates an object used to perform LocallyLinearEmbedding based dimensionality reduction	uses `n_neighbors` neighbors for each point and `n_components` coordinates for the manifold
`manifold.MDS(n_components)`	Creates an object used to perform MDS based dimensionality reduction	immerses the dissimilarities in `n_components` dimensions
`manifold.TSNE` `(n_components,perplexity)`	Creates an object used to perform *t*-SNE based dimensionality reduction	uses `perplexity` nearest neighbors to embed into `n_components` in manifold learning algorithms
`manifold.Spectral Embedding` `(n_components)`	Creates an object used to perform Spectral embedding for nonlinear dimensionality reduction	projects into a subspace with `n_components`
`decomposition.PCA` `(n_components)`	Creates an object used to perform PCA	keeps `n_components`
`decomposition.KernelPCA` `(n_components,kernel)`	Creates an object used to perform KernelPCA	keeps `n_components` after using a `kernel` type
`decomposition.Truncated` `SVD(n_components)`	Creates an object used to perform dimensionality reduction using truncated SVD	keeps `n_components` in the output data
`decomposition.non_negative_` `factorization(X)`	Creates an object used to perform NMF	`X` is the input data
`random_projection.` `GaussianRandomProjection` `(n_components)`	Creates an object used to perform dimensionality reduction through random projections	projects onto `n_components` in the target space

Table 11.34 Handy commands in sklearn for supervised and unsupervised Machine learning methods and evaluation

Command	Action	Notes
metrics.roc_auc_score (y_true, y_score)	Computes the Area Under the Receiver Operating Characteristic Curve (ROC AUC) from prediction scores	y_true are true labels y_score are predicted
metrics.f1_score(y_true, y_pred)	Computes the f1-score of the predicted labels	y_true are true labels y_pred are predicted
metrics.mean_squared_error(y_true, y_pred)	Computes the Mean squared error regression loss	y_true are true values y_pred are predicted
metrics.silhouette_score(X, labels)	Computes the mean Silhouette Coefficient of all samples	X is an array of pairwise distances between samples; labels are their cluster labels
model_selection. cross_val_score (estimator, X, Y, cv)	Performs cross-validation	estimator is the classifier to be evaluated, X is the data, y is the set of labels and cv is the number of folds
tree.DecisionTree Classifier(criterion)	Creates a decision tree classifier	criterion is the splitting criterion
naive_bayes.MultinomialNB(alpha)	Creates a Naïve Bayes Classifier	alpha is the Laplacian smoothing parameter
neighbors.KNeighbors Classifier(n_neighbor)	Creates a Nearest Neighbor Classifier	considers n_neighbor neighbors
ensemble.Bagging Classifier (base_estimator, n_estimators)	Creates a Bagging Ensemble	base_estimator uses n_estimators in the ensemble
ensemble.AdaBoost Classifier (base_estimator, n_estimators)	Creates an Adaboost classifier	base_estimator is the base classifier in the ensemble and n_estimators is the number of classifiers in the ensemble
svm.SVC(C, kernel)	Creates an SVM classifier	C is the regularization parameter with a kernel type
neural_network.MLPClassifier (hidden_layer_sizes, activation)	Creates a neural network classifier	hidden_layer_sizes is the number of neurons in the hidden layers and activation is the activation function
cluster.KMeans(n_clusters)	Creates a K-Means clustering object	generates n_clusters clusters
mixture.Gaussian Mixture(n_components)	Creates a Gaussian Mixture Model	uses n_components in the mixture

References

1. Agrawal, M., Kayal, N., & Saxena, N. (2004). Primes is in **P**. *Annals of Mathematics, 160*(2), 781–793.
2. Ahmadi, M., Ulyanov, D., Semenov, S., Trofimov, M., & Giacinto, G. (2016). Novel feature extraction, selection and fusion for effective malware family classification. In *Proceedings of the Sixth ACM Conference on Data and Application Security and Privacy* (pp. 183–194)
3. Anonymous (2019). *Tychonis Brahe Dani Opera Omnia*. London: Forgotten Books.
4. Azzalini, A., & Bowman, A. W. (1990). A look at some data on the old faithful geyser. *Applied Statistics, 39*, 357–365.
5. Devlin, D. (1991). *Logic and information*. Cambridge: Cambridge University Press.
6. Ekman, P. (2008). An argument for basic emotions. In *Basic emotions, rationality, and folk theory* (pp. 169–200). New York: Springer.
7. Fisher, R. A. (1936). The use of multiple measurements in taxonomic problems. *Annals of Eugenics, 7*(2), 179–188 (1936)
8. Fleiss, J. L. (1981). *Statistical Methods for Rates and Proportions*. London: Wiley.
9. Galois, E. (1830). Sur la theorie des nombres. *Bulletin des Sciences Mathematiques, XIII*, 428.
10. Guyon, I. (2003). Design of experiments of the nips 2003 variable selection benchmark. In *NIPS 2003 Workshop on Feature Extraction and Feature Selection* (Vol. 253).
11. Hopcroft, J., & Ullman, J. (2000). *Introduction to Automata Theory Languages and Computation*. Cambridge: Cambridge University Press.
12. Johnson, R. W. (1996). Fitting percentage of body fat to simple body measurements. *Journal of Statistics Education, 4*(1), 265–266.
13. Kanade, T., Cohn, J. F., & Tian, Y. (2000). Comprehensive database for facial expression analysis. In *Proceedings of the 4th IEEE Conference on Automatic Face and Gesture Recognition* (pp. 46–53)
14. Krizhevsky, A., Sutskever, I., & Hinton, G. E. (2012). Imagenet classification with deep convolutional neural networks. In F. Pereira, C. J. C. Burges, L. Bottou, & K. Q. Weinberger (Eds.), *Advances in neural information processing Systems* (Vol. 25). Red Hook: Curran Associates, Inc.
15. LeCun, Y. (2010). MNIST handwritten digit database. http://yann.lecun.com/exdb/mnist
16. Mainali, S., Garzon, M., Venugopal, D., Jana, K., Yang, C.-C., Kumar, N., Bowman, D., & Deng, L.-Y. (2021). An information-theoretic approach to dimensionality reduction in data science. *International Journal of Data Science and Analytics, 12*, 1–19.
17. Miller, G. (1995). Wordnet: A lexical database for english. *Communications of the ACM, 38*, 39–41
18. Mitchell, T. M. (1997). *Machine Learning*. (McGraw-Hill, New York, 1997)
19. Ronen, R., Radu, M., Feuerstein, C., Yom-Tov, E., & Ahmadi, M. (2018). Microsoft malware classification challenge (2018). CoRR abs/1802.10135
20. Strang, G. (2016). *Introduction to Linear Algebra*. Wellesley: Wellesley-Cambridge Press.
21. Turing, A. M. (1936). On computable numbers with an application to the entscheidungsproblem. *Proceedings of the London Mathematical Society, 42*(2), 230–265. A correction, ibid 43: 544–546, 1936.
22. Yang, X., Garzon, M. H., & Nolen, M. (1960). A coefficient of agreement for nominal scales. *Educational and Psychological Measurement, 20*, 37–46.

Printed in the United States
by Baker & Taylor Publisher Services